The African Savanna

MANAGING EDITORS
Amy Bauman
Barbara J. Behm

CONTENT EDITORS
Amanda Barrickman
James I. Clark
Patricia Lantier
Charles P. Milne, Jr.
Katherine C. Noonan
Christine Snyder
Gary Turbak
William M. Vogt
Denise A. Wenger
Harold L. Willis
John Wolf

ASSISTANT EDITORS
Ann Angel
Michelle Dambeck
Barbara Murray
Renee Prink
Andrea J. Schneider

INDEXER
James I. Clark

ART/PRODUCTION
Suzanne Beck, Art Director
Andrew Rupniewski, Production Manager
Eileen Rickey, Typesetter

Copyright © 1992 Steck-Vaughn Company

Copyright © 1989 Raintree Publishers Limited Partnership for the English language edition.

Original text, photographs and illustrations copyright © 1985 Edizioni Vinicio de Lorentiis/Debate-Itaca.

All rights reserved. No part of the material protected by this copyright may be reproduced or utilized in any form by any means, electronic or mechanical, including photocopying, recording, or by any information storage and retrieval system, without permission in writing from the copyright owner. Requests for permission to make copies of any part of the work should be mailed to: Copyright Permissions, Steck-Vaughn Company, P.O. Box 26015, Austin, TX 78755. Printed in the United States of America.

Library of Congress Number: 88-18337

2 3 4 5 6 7 8 9 0   97 96 95 94 93 92

Library of Congress Cataloging-in-Publication Data

Beani, Laura, 1955-
 [Savane africane. English]
 The African Savanna / Laura Beani, Francesco Dessi.

 — (World nature encyclopedia)
 Translation of: Savane Africane.
 Includes index.
 Summary: Describes the geographical features, climate and plants and animals of the African grassland with emphasis on their interrelationship.
 1. Savanna ecology—Africa—Juvenile literature.
2. Biotic communities—Africa—Juvenile literature. [1. Savanna ecology—Africa. 2. Biotic communities—Africa. 3. Grasslands. 4. Ecology—Africa.]
 I. Dessi, Francesco. II. Title. III. Series: Natura nel mondo. English.
QH194.B4313   1988   574.5'.2643'096—dc19   88-18363
ISBN 0-8172-3325-3

WORLD NATURE ENCYCLOPEDIA

The African Savanna

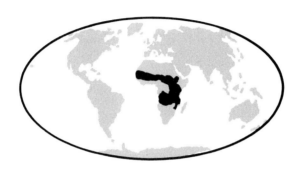

Laura Beani
Francesco Dessi

Laramie Junior High
1355 N 22nd
Laramie, WY 82070

Austin, Texas

CONTENTS

6 INTRODUCTION

9 THE WORLD OF THE SAVANNA

Features and Frontiers of the Masai Mara-Serengeti Savanna, 9. The Landscape, 11. The Nature and History of the Terrain, 11. The Rains and the Dry Season, 13. The Grasslands and Their Trees, 16. Savanna Fires, 18. The Dynamics of the Serengeti, 22. A Glimpse at the World of the Termites' Nest, 25.

29 GRASSLANDS AND HERBIVORES

The Effects of Grazing, 29. Grass: An Ideal Food? 30. The Feeding Tactics of Herbivores, 33. The Impala, the Topi, and the Buffalo: A Case of Ecological Separation, 35. The Large Migratory Animals, 37. The Elephant Problem, 39.

43 HERBIVORE SOCIETY

How Food Availability Affects the Social Structure: The Case of the Antelope, 43. The Group as a Defense, 47. The Harem, the Family, and Solitary Life: Other Herbivore Behavior, 51. The Mating Season, 56.

61 CARNIVORES

The Predator-Prey System, 61. Hunters or Scavengers? 63. Hunted Prey and Hunting Techniques, 65. Social Behavior of The Carnivores: The Group vs. Solitary Life, 74.

83 PRIMATES

Origins, 83. Communication Among Guenon Monkeys, 85. Cooperation Among Baboons, 87.

91 BIRDS AND REPTILES

The Savanna Birds, 91. Landlubbers: Ostriches and Bustards, 93. Nesting Problems: The Weaverbirds, 95. Birds of Prey: Scavengers and Hunters, 98. A Word About Reptiles, 102.

107 GUIDE TO AREAS OF NATURAL INTEREST

Senegal, Ivory Coast, Ghana, Nigeria, Cameroon, Sudan, 108. Uganda, 110. Zaire, 114. Kenya, 115. Tanzania, 118. Zambia, 122. Mozambique, 122. Zimbabwe, 122.

123 GLOSSARY

126 INDEX

128 PHOTO CREDITS

INTRODUCTION

The word *savanna* brings to mind the image of a vast expanse of grassland. The expanse is dotted with trees and stretches as far as the eye can see. It is a tropical landscape that is easy to name but not easy to describe. The savanna lands of Africa, South America, and Australia share certain common features. But the natural systems in which they exist differ. There may be several types of savanna on one continent. In Africa, for example, the plains of Guinea form a savanna which is thickly carpeted with palm trees. This savanna is quite different from the savanna grasslands of Kenya, Tanzania, or South Africa.

In the savanna, many distinct, self-contained habitats called "microhabitats" exist. A savanna may contain hill country inhabited by elephants. Scraggy trees, scattered to the farthest horizon, may serve as an eagle's perch. There may be green slopes grazed by antelope. Or there may be low-growing scrublands, called the "bush," hiding groups of dik diks, which are miniature antelope. Among trees growing beside a stream or river, there may be slow, gangling-gaited giraffe. Storks perch in marabou trees. Clumps of euphorbia trees, shaped like candelabras, are filled with

guenon monkeys. During the heat of the day, acacia trees shade gazelles beneath a canopy of branches. Kopjes, which are large, lone outcrops of rock, stand above the grass and provide homes for lions.

The Masai Mara-Serengeti is a vast protected area measuring 9,650 square miles (25,000 square kilometers). The two parks called Masai Mara-Serengeti form a savanna ecosystem. Physically, the parks adjoin each other. Politically, they are divided by the boundary between Kenya and Tanzania. Together they form one of the world's last and most important concentrations of large mammals left in the wild. Approximately thirty species of hoofed mammals, or ungulates, and thirteen species of flesh-eating animals, or carnivores, are found here.

For at least one million years, the Mara-Serengeti ecosystem has remained almost unchanged. But this does not mean that the area may not undergo changes brought on by a range of factors. Drought, plagues, population increases, and the spread of cultivated lands along the park's boundaries all cause changes to occur. The savanna is part of history and time. It has its own dynamic and complex balance.

THE WORLD OF THE SAVANNA

The savanna with its wooded grasslands and animal life forms what is known as an "ecosystem." An ecosystem is an association of plants and animals which interact together in a particular physical environment. The savanna lies between equatorial rain forest on one side, and arid, spiny scrublands on the other. So it is wedged between two very different environments. One is lush; the other is parched. Can it be explained as a system that is stable and well defined? Or is it an environment that is half forest and half scrub? These are unanswered questions. It is hard to tell the difference between the savanna and similar neighboring zones that are also covered with scrub and sparse trees. Also, map boundaries claiming to outline the savanna may not reflect true limits.

Features and Frontiers of the Masai Mara-Serengeti Savanna

A savanna can most simply be defined as a tropical ecosystem. Its covering of grasses and scattered trees and bushes are capable of surviving periods of drought. This definition is general enough to apply to the Masai Mara-Serengeti ecosystem. Perhaps the most striking feature of this ecosystem is the variability of the environment. Physically, the landscape is varied. It ranges from prairielike lowlands to wooded areas with a cover of thick grass. Each year has its alternating rainy and dry seasons. Animal migrations are seasonal. During migrations, herds of wildebeest or gnu form long columns numbering in the thousands. They migrate in search of new grass on which to feed. Their movements trace boundaries of the savanna that have existed for a long time.

The land formations around the edges of the Masai Mara-Serengeti halt movements of these large herds of herbivores. To the north lie the Loita Plains. They are dry and inhospitable when migrating animals arrive. To the east rises Ngorongoro Mountain. It is an extinct volcano whose large crater now forms an ecosystem of its own. Passing herds of animals come close but they do not enter Ngorongoro Mountain. To the south, the boundary of the Serengeti forms a loop to include the small alkali lake of Lake Legaja or Lake Ndutu. In the local tongue, the names mean "place of peace, dedicated to God and silence." Beyond the crater, farther south, dense, dry forests line the slopes around Lake Eyasi. Farther on, the farmland crops of the Sukuma tribe spread to the west. They mark the edges of the park. It is

Preceding page: An elephant has a dust bath in Kenya. When traveling with four-wheel drive it is easy to find elephants even outside the parks. They are attracted to streams and water holes where they bathe. They spray water, mud, sand, or dust all over themselves with their trunks to cool off and get rid of insects. Elephants are always on the move, especially at night, due to their food and water requirements.

Opposite page: A typical savanna view in the Serengeti is pictured during the rainy season. An *Acacia tortilis* tree stands in the foreground and grass and scattered trees stretch as far as the eye can see. The term *savanna* most probably derives from "savannah," a word used by North American Indians to describe the prairies. Today, a distinction is made between temperate grasslands and tropical and subtropical savanna. But they are all grouped as "African grasslands."

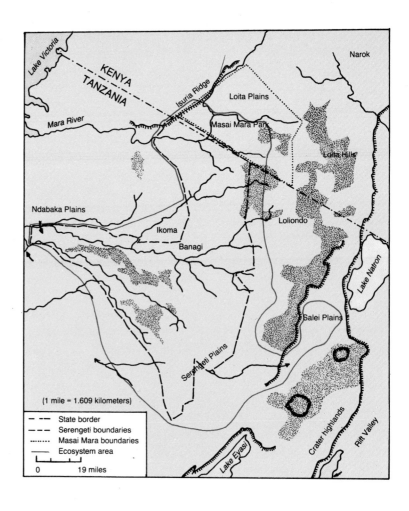

The Masai Mara-Serengeti ecosystem is the area where wildebeest migrations take place. It partly overlaps the park areas. The shaded areas on the map indicate hills. On the whole, the Masai Mara reserve covers 700 sq. miles (1,812 sq. km), the Serengeti Park covers 5,600 sq. miles (14,500 sq. km), and the Ngorongoro Conservation Area covers 2,500 sq. miles (6,475 sq. km.). The conservation area adjoins the Serengeti and is skirted by the wildebeests' migratory routes. Such large areas make it possible to study grass-herbivore relationships and their interactions with predators. The confined but wide ecosystem allows a natural balance to develop. The area is an extraordinary world in miniature.

possible that the rise in the elephant population over the last twenty years is due to the slow but steady advance of these cultivated areas. This is forcing animals ever deeper into the game reserve.

Zebra, wildebeest, and herds of Thomson's and Grant's gazelles roam the park. During migration, they make their way from the rich, flat grazing grounds of the south toward the western corridor. They start their journey at the end of the rainy season when the grass starts to dry. They reach the northern grasslands at the end of the dry season. When the rains return, their journey starts all over again, in reverse. Herds that scatter across the scrublands in the north, staying close to permanent water sources, join ranks. Gradually, they form a series of close-knit columns. Slowly, they head back to the low grasses of the southern plains. Here their young are born, and here they will stay until the rains are over.

A mixed herd of buffalo and wildebeest grazes in Tsavo Park, Kenya. In the grass to the right, there are crowned cranes. In the midst of the buffalo are quite a few cattle egrets. Such egrets often mingle with herds in search of insects. It is quite common to find these two species of herbivores grazing together and sharing watering holes. Competition between them does exist, however, especially during the dry season. During the 1960s, the buffalo population grew rapidly from thirty thousand to fifty thousand head. This growth followed the control of cattle plague. But since the 1960s, growth has been limited by the explosion of the wildebeest population, which is still on the rise.

The Landscape

At first glance the savanna may not appear to have a dense animal population. But the Masai Mara-Serengeti area is said to have two and a half million hoofed mammals. At times, the plains of the savanna are so thick with thousands of animals that they appear black. Animals are a fundamental part of this landscape, even if they are always on the move. When herds are grazing, they look like blurred shapes, barely moving against the landscape.

The savanna horizon is boundless and apparently empty, with just a scattering of trees. It stretches forever, to the highlands of the Rift Valley or the rugged bluff of a lake or volcano. Fritz Jaeger was one of the first Europeans to cross the high plateaus of East Africa. In his 1907 diary he wrote about gazing out upon the savanna and seeing only endless grass and sky. The American writer Ernest Hemingway also spent a great deal of time in the African savanna. In his book *Green Hills of Africa*, he felt as if he were in Spain as he crossed the splendid landscapes.

The Nature and History of the Terrain

The rolling Serengeti region is an upland plateau or tableland. It ranges from a height of 6,900 feet (2,100 meters) in the east, where it borders the Rift Valley, to 4,165 feet (1,270 m) around Lake Victoria.

The ridge from which Hemingway viewed the lake is

A flock of flamingos wades in the alkali lake in the Ngorongoro Crater. Large flocks of flamingos are very frequent visitors to the alkali lakes of East and Central Africa, along the Rift Valley. At Lake Nakuru, in Kenya, flocks number millions of birds. The lesser flamingo, in the foreground, is smaller, with bright and glossy pink plumage. The greater flamingo is larger in size. It has white plumage, with black wing quills and coral red coverts.

the edge of the Rift Valley, which fringes the Ngorongoro Plateau. Ngorongoro is a collapsed volcano. It once rose to a height of 16,400 feet (5,000 m). About two million years ago, it exploded and covered thousands of square miles with a thick layer of ash. What remains today is a very large crater, a caldera, more than 2,300 feet (700 m) deep. It has a diameter of about 12 miles (20 km). At the rim, it reaches a height of 6,500 feet (2,000 m). It is a basin lush in green vegetation with a salt lake in the middle. It is densely populated with a variety of mainly nonmigratory mammals.

The low grass of the southeastern Serengeti plain is to some extent the product of the volcanic nature of the soil. About 40 inches (1 m) below the surface, a continuous, impermeable crust has formed. It is rich in calcium carbon-

ate, formed from the weathering activity of rain on volcanic rubble. This crust stops deep roots from working their way downward. Only small plants with shallow root systems can survive here. They must be able to live with high salt and alkali levels in the soil. This soil contains large amounts of potassium, sodium, and calcium, with smaller quantities of magnesium.

To the west, layers of volcanic rocks date back to the Precambrian time, which began some two and one-half billion years ago. They are covered by more recent sedimentary deposits. Across time, most of the alkali and salt eroded from these soils. Thus, the soil is rich and deep, and the cover, "sea of grass" above it, stands 3 to 6 feet (1 to 2 m) tall. Roots extend deeply through a penetrable layer of pebbles, and the grass grows thick.

A mountain-forming event that occurred in the Precambrian times shaped the hills and highlands on the eastern boundary. These are desolate mountains. Their quartzite and gneiss slopes are covered with acacia bushes and thorny scrubs. Granite outcrops, or kopjes, are also from the same period. The amount of humus in the ground determines whether a kopje is bare or covered with trees or grass. Sometimes humus collects in small pockets formed by cracks in the rocks. Therefore, each kopje develops a particular world-in-miniature, with its own animal life.

The Rains and the Dry Season

In the savanna, American zoologist George Beals Schaller noted that the rains "arrive and depart like a flower blossoming." They follow the course of the monsoons. These winds originate in the Indian Ocean. They blow from the northeast for six months a year. For the rest of the year, they blow from the southeast. With them, they bring a great deal of rain. Generally, between November and June, there is enough moisture for grass to grow. On the savanna, grass sprouts in just one day. But between July and October, the ground dries out. During these months, the grass grows more slowly and sparsely. Heavy rains fall in March, April, and May. Light rains fall in November and December. It is hard to define a typical year, for the rains vary in how early or late they arrive.

Rain is the beat for the vital rhythm of the changing landscape of the savanna. With the arrival of the dry season, the ungulates, or hoofed mammals, migrate. They move from the low-lying grasslands of the south into northern valleys with tall grass cover. The end of the rainy season comes gradually. By mid-March, the long grasses turn to a lighter green and then take on a hint of yellow. Haze starts to shroud the hills. It is almost time to head for the grazing grounds in the northwest. In this period, wildebeest reach the peak of their mating season. Like the nomadic herdsmen of the bush and the steppe, they seem to know where new grass is growing. Herdsmen and wildebeest pursue the rare clouds in the sky for the hidden signs of rain. When the rainfall in November is very light, the journey back to the southern grasslands is postponed. The herds wait in the wooded regions of the north.

A herd of wildebeest fords the Mara River, in the Masai Mara Reserve. Wildebeest herds travel on long routes, sometimes crossing deep rivers. This can mean death to many individuals, especially the young. They move in the endless search for green pastures where the grass is in the growth stage. They eat almost any kind of grass, provided it is in the right stage of growth. An analysis of stomach contents during the dry season shows that they pick the greener and most tender parts of grasses. They discard tough leaves and stalks. This means that they must alternate their grazing grounds and lead a nomadic life. This photo was taken in August, during the dry season, when the large herds move toward the northern pasture lands.

Drought is an event which changes the course of nature. Events which occur in parks are natural events, but they can be influenced in many ways. In 1967, the growth period for the grasses was limited to 83 days in the bush area close to Banagi (as opposed to 198 days in 1968) and to 56 days in the lowlands at the foot of the Naabi Hills. The effect on the hoofed mammal population was enormous.

In the following years, the trend was reversed. It rained even in the dry season. In the wooded regions of the north, the rainfall in the dry months increased from 6 inches (150 millimeters) to 10 inches (250 mm) between 1971 and 1976. There was new grass all year round. In 1977, the wildebeest

savanna with scrub

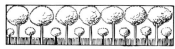
scrub or bush

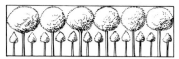
arid-deciduous forest (leaf-shedding trees)

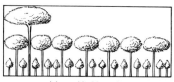
mainly semideciduous forest

mainly evergreen forest

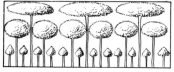
rain forest

The vegetation sequence in central-western Africa ranges from savanna to rain forest. The sequence is based on an increasing level of humidity. A forest is defined as a rain forest *(bottom picture)* when it has three complete layers of trees. In the evergreen and semideciduous forest, the upper layer is continuous. In the arid-deciduous forest, the upper layer is not present at all. In the scrub or bush, there is just one tree layer and a moderate grass cover. In the savanna *(top picture)*, if wooded, grass is dominant. The height of the savanna grass cover, while in the growth phase, is never less than about 32 inches (80 cm).

reached the incredible total of 1,300,000. They numbered five times their population in the sixties. But this did not cause a drop in the amount of food available per head. The arid lowlands, once roamed by a handful of oryx, Grant's gazelle, and the occasional ostrich, became crowded with medium and large herbivores. The kongoni hartebeest, topi antelope, and buffalo were joined by the waterbuck or kob. This was a sure sign of the high level of moisture in the ground.

The Grasslands and Their Trees

A "sea of grass" is not necessarily an even surface. It changes with differing types of soil, the degree of humidity, and the type of grazing. It features distinctive plant associations. These associations tell the geologic history of the grassland. In the eastern plains, the terrain has high salt and alkali levels due to recent volcanic activity. The grasslands are low and broken up. Erosion occurs frequently. Erosion is accelerated sometimes by the presence of huge numbers of herbivores.

Small plants with short roots are common. They grow along the ground. In arid areas, these plants tell how heavily the grassland is grazed. Most of the grasses belong to the genus *Aristida,* also called "three-bearded grass" because of the appearance of their seeds. After a short rain shower, these plants sprout all at once on the ground. But animals do not feed on them. As a result, the plants begin to dominate areas that have been overgrazed. Where they dominate, the savanna gradually starts to turn into desert.

The real sea of grass is the high grassland of the wooded regions and pasture lands to the north. Here species grow to a height of more than 6 feet (2 m). The wind makes these tall grasses sway back and forth. This constantly changes the shades of color on the landscape. Here the tallest forms of grasses are found. In humid areas around seasonal marshes, *Pennisetum purpureum,* also called "elephant grass," grows. It grows with Guinea grass or *Panicum maximum,* which can reach a height of 13 feet (4 m). This is buffalo and elephant country. But it is also roamed by many small species of antelope which feed on leaves and shoots. Clearings filled with *Themeda triandra* are common. This red type of oat grass stands up well to fire.

There was a time when forests covered the area now occupied by *Themeda.* But fires open broad gaps between the trees. Grass of the genus *Hyparrhenia* also stands up

A group of eland antelope wanders past a row of baobab trees. The baobab tree has a huge trunk which is swollen and spongy. It is a rich reserve of water. It has only sparse leaves. This limits transpiration. One recurrent feature in the trees and shrubs of the savanna is the thickening of the bark. This makes them fire resistant. They also have very well-developed root systems. The far-reaching roots grow just 4 to 8 inches (10 to 20 cm) below the surface. They absorb as much water as possible, even when the rain barely dampens the ground.

well to fire. It actually benefits from periodic fires. Only the new leaves of *Hyparrhenia* are edible. So if herbivores do not graze the area continually, this species will grow to a height of over 13 feet (4 m). When grasses grow that tall, clearings become thick with vegetation. Animals find it difficult to enter and graze. So they avoid such clearings.

The variety of plant forms that grows on the grasslands and the patchy distribution of rainfall give the traveler the impression of passing through different seasons. Across the grassland, spring green blotches appear in the midst of the late-summer yellows. Animals gather by the thousands in open spaces. They move beyond a hill and disappear into tall green grass. They follow a random but logical pattern and prefer certain foods to others. They rotate their grazing areas. In this way, they keep the grass growing.

These pages picture, from left to right, a series of savanna grasses of the Graminaceae family. The genus *Aristida* is common in climates with low humidity, especially semi-arid and arid zones. It is not a popular food. It represents one of the dominant species in overgrazed areas. Many species of the genus *Andropogon* are grazing-dependent. They benefit from being grazed frequently. *Pennisetum purpureum* is a very sought-after species. It is common in wooded savanna, as is *Panicum maximum*. The genera *Themeda* and *Hyparrhenia* are fire-resistant and dominate areas where fire is frequent.

The balance is upset when, for various reasons, too many animals concentrate in an area. As he crossed the Masai lands, heavily grazed by domestic livestock, Hemingway wrote about the discouraging look of the plains. This landscape is typical of the moister parts of the savanna. Here large trees with big leaves grow with a tall, thick species of grass called *Loudetia karagensis*. They are remains of ancient forests, now gone.

There is no gradual change from the forest environment to the grassland environment. This is because fires continually remove underbrush and young plants. Fires also make the age and size of trees uniform, creating the illusion of a deer park.

In grassland regions with spiny or thorny forests, the genus *Acacia* is dominant. Thirty-eight species of trees belong to this genus. Ten species of Acacia trees are found in Africa's grassland areas. They account for 45 percent of the trees. The most common acacia tree is the small *A. drepanolobium*, which is quite fireproof. A taller form, *A. clavigera*, grows on hilly slopes. At the edges of grasslands, *A. tortilis* grows. It has a distinctive umbrellalike canopy of foliage.

Savanna Fires

Fire affects the landscape of the savanna. It leaves its mark wherever it passes. By comparison, the effect of graz-

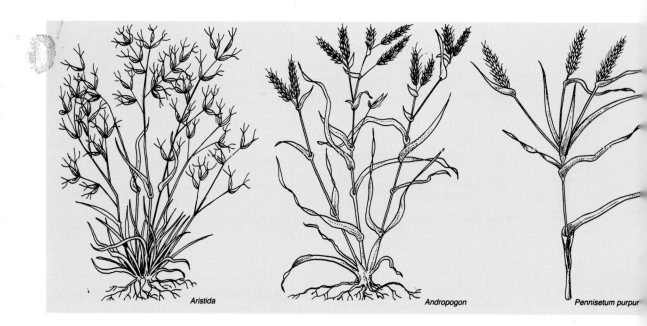

Aristida   Andropogon   Pennisetum purpur

ing animals is less dramatic. Almost every year, about two-thirds of the bush and much of the grasslands burn. Fires are started by herdsmen and farmers on the boundaries of the park. Inside the park, they may be started by poachers trying to wipe out evidence of their work. Or they may be started by honey-gatherers, who use smoke to flush out swarms of bees. Lightning can also start fires, but this is less common since violent downpours usually accompany lightning.

Fire is not necessarily a catastrophe. It occurs again and again and is a natural event during every dry season. The livestock-tending tribes of Africa are certainly well aware of how fire can be properly used. Like water, fire comes in many forms and has a varying impact on the environment.

At the start of the dry season, the grass still contains a high percentage of water. At this time, a fire simply consumes the layer of leaves and dead grass on the ground. As long as the pile-up of flammable material is not too big, the fire will not burn far. If a grassland area burns every so often, there will not be much accumulation of flammable material. This slow, low-burning, "cool" type of fire does not attack trees. Because there is no rise in ground temperature half an inch below ground level, neither seeds nor roots burn. Animals in their lairs and burrows are not affected either.

But by July, fires become more frequent, widespread

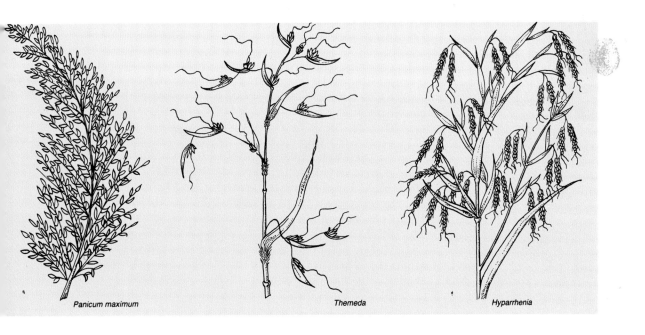

Panicum maximum *Themeda* *Hyparrhenia*

An acacia tree grows in Amboseli National Park, Kenya, in a savanna area with a thick covering of shrubs. Hanging from the branches of the tree are nests built by weaverbirds. In the African savanna, these birds gather in groups. They nest close to one another, forming easily visible colonies.

savanna with scrub or bush

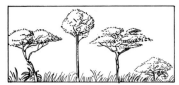
savanna with scattered trees and shrubs

savanna with just shrubs

savanna with just grass

There are at least four distinct types of savanna. In savanna with scrub or bush, trees and bushes form a continuous but thin cover. Some savanna has scattered trees and shrubs. Other savanna has just shrubs and bushes. Savanna with just grass has no trees or bushes. Repeated fires turned areas once covered by forests into wooded savanna or grassland savanna during the last few centuries. This change is due to the increase in the human population.

and destructive. Immature trees and shoots are vulnerable. Plant life under 3 feet (1 m) is 90 percent destroyed. Plants up to 6 feet (2 m) are 70 percent destroyed. But large trees are barely touched, for in the Serengeti, a fire rarely reaches the treetops. As a result, it is the growth of young trees and the spread of thickets and scrub that are slowed down by fire.

Fire clears underbrush and dead trees from the grassland. It is also responsible for the regular spacing of the vegetation that remains. Often this vegetation seems to be laid out like a garden. Forests exist where all the trees are the same age. This is because fire may not visit an area for seven or eight years. So trees tend to be grouped depending on these intervals. In the bush, fire follows tracks opened by animals, or it follows routes taken by previous fires. It attacks leaves and devours old trunks. It spares acacia trees, which stand up well to fire once they are over three years old. Fire clears wide open spaces. Without fires, much of the Serengeti would be invaded by the bush in just a few years.

The Dynamics of the Serengeti

Rain, fire, grass, and animals are all main characters in the always changing Masai Mara-Serengeti ecosystem. Interaction between these elements is so complex that it is difficult to discuss one of them individually. During the seventies, rainfall increased during the dry season. The rain gave rise to a richer stand of grass. This resulted in an increase in the population of herbivores. They spread into areas once too arid to graze on.

In these years, the numbers of hoofed mammals also increased for another reason. Cattle plague was eliminated from the area. The plague used to start within the livestock population, and then spread to wildlife. During the sixties, livestock owned by Masai farmers was systematically vaccinated against plague. In 1962, the last outbreak of the plague called "murrain" occurred among the wildebeest. In 1963, there was a final outbreak among the buffalo. Between 1961 and 1967, both species doubled their numbers, but there was no increase among zebras. Zebras are not ruminants, or cud-chewing animals, and therefore are not susceptible to the plague virus.

During the seventies, there was also a decrease in the frequency and size of fires. The frequency of fires was reduced, in part, because of heavier grazing by wildebeests. By feeding on tall, dry grasses, wildebeests remove flammable material which fires feed on in the dry season. In addi-

Masai herdsmen tend their cattle. In overgrazed areas, there are patches of bare ground and clumps of grasses on which the cattle do not graze. This increases the erosion process, and the land gradually turns into desert. As a rule, the large carnivores do not attack livestock, if they can choose other prey. Wild hoofed mammals keep their distance from cultivated land, with the exception of the elephant and rhinoceros. The greatest interactions between wildlife and cattle involve disease and overuse of soil. In 1890, an epidemic of cattle plague, probably imported from Asia, destroyed about 95 percent of the domestic livestock and the populations of buffalo and wildebeest.

tion, intensive grazing in these grasslands creates areas where there is not a blade of grass. Wildebeests can remove 85 percent of the grass cover from an area in a few hours. Fires cannot spread in such surroundings. The result is a speedup in the regrowth of acacias and small trees. This, in turn, causes an increase in the number of giraffes and other herbivores which feed on leaves and shoots. At the same time, the number of trees reaching maturity drops because of the feeding habits of giraffes and elephants. These large herbivores are being forced into the park by the ever-increasing number of cultivated lands along its boundaries.

The numbers of hyenas and lions, which mainly hunt wildebeests, are also increasing. But the African hunting dog is on the verge of extinction. The hunting dog's population dropped from about one hundred in 1960 to about thirty in 1977. Competition for food may be the reason for the drop. But the fact that hyenas prey on the hunting dog's young may also contribute. The cheetah is less vulnerable because it is diurnal. It hunts during the day when other species are not active. It also devours its victims without delay. The leopard also is not affected by competition because it drags its prey into trees where hyenas and lions do not venture.

This diagram shows interrelations in the Masai Mara-Serengeti ecosystem during the 1970s. Various members of the ecosystem responded to the increase in rain during the dry season which occurred in the 1970s. A savanna which is in balance shows no signs of desert formation. Balance means that grazing is not too heavy. Grazing areas are properly rotated and herbivores are in balance with their predators. Balance between animals depends, in part, on the availability of prey. In the 1970s, the lion population on the Serengeti plains almost doubled. The population of hyenas rose by 50 percent. These increases are amazing in view of the fact that these carnivores are territorial. Their growth rate is not directly related to the numbers of nomadic herbivores.

This web of interacting creatures forms a dynamic system of life on the African plain. Yet this is a simplified view. In reality, there are many more interactions at every level of the food chain. Competition plays its role, and many side effects depend on the outcome of competition. The effects may be short term or long term. For example, intensive grazing by the wildebeest may lead to the spread of some grasses of the Graminaceae family which have protected growth buds. Such grass species prevail over others with more vulnerable tops, which are usually preferred by more selective hoofed mammals. During the grazing season, for example, Thomson's gazelles follow the tracks of the wildebeest. By following them, they are sure of finding grass which is regrowing and still tender.

Population increase of a species does not go on forever. At a certain point, there is a shortage of food resources. At times, predators increase in number. Predator growth rate is directly related to prey growth rate. An epidemic disease may halt the population. Wildebeests interact with various species of herbivores, predators, and other wildebeests. Even species of grasses compete with each other for light, water, and room to grow.

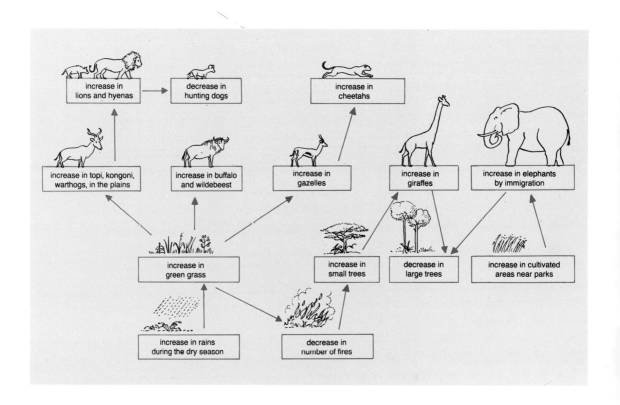

Soon after sunset, the dung beetles arrive. They fly in large buzzing swarms. It is almost an attack because in fifteen minutes more than 3,800 beetles can gather on an elephant dropping. Immediately they begin to dig tunnels. In half an hour, they will reduce the pile of dung to a wide "carpet" covered with a thin layer of fibrous material formed by undigested plant parts. In the drawings, a female dung beetle is digging around some dung to make it into a ball. She shapes it, then rolls it to the place where it will be buried. Typically, she stands on her front legs and pushes with her hind legs. Then she buries the ball of dung and lays a single egg in it. The work of the beetles aerates the soil and makes it finer. It delays the spread of parasites and disease-causing microbes while reducing the number of flies.

The savanna is densely populated by a variety of creatures which take advantage of rotting and decomposition processes. These creatures make up a part of the large biomass for the area. The *biomass* is the total weight of all living organisms in a given area. Termites and certain beetles, play an important ecological role. They remove the droppings of larger mammals. If these droppings were not removed, they would suffocate plant life. Insects who work to remove droppings have adapted to their environments. They are specialized, that is, they have adapted to the special conditions of their environments.

A Glimpse at the World of the Termites' Nest

The landscape of the savanna is distinctively marked by the silent but important work of termites. Humivorous, or humus-eating, termites feed on decomposing organic matter. Lignivorous, or wood-eating, termites attack plants and wood.

Small dirt mounds rise up here and there from the floor of the savanna. They are the work of the humus-eating

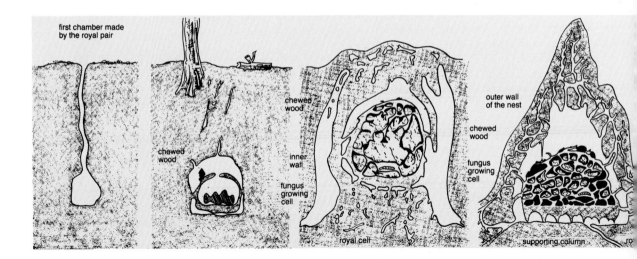

Above: The construction and structure of the nest of the fungus-cultivating African termite *Bellicositermes natalensis* is shown. The outermost chambers are filled with fungi and chewed wood. There the fungi grows. Air circulation is allowed around the fungus nursery. The inner areas house eggs, larvae, workers, soldiers, and the royal cell, where the founders of the colony live. The couple consists of a tiny king and a queen with her grotesquely swollen body. Her body is an "egg-laying machine." To the right in the drawing is a giant pangolin or anteater, a common predator of the termites' nest. Termites, together with ants, form an essential part of the anteater's diet.

Right: Termites nest near Lake Rudolf in Kenya. The columnlike structure assures air circulation and "air-conditions" the nest. Inside the nest, the humidity and temperature are relatively constant. The nest is a suitable refuge for many vertebrates and invertebrates in the tropics.

termites. In central-western Africa, these mounds take the shape of large mushrooms. These mounds are protected by clay roofs. The roofs are laid one on top of the other to protect the nest against rain. In the forests of Tanzania, the termites account for 80 percent of the biomass of the animals found within the soil. The termites alone make up half of this population.

Among the wood-eating species, the macro-termites stand out. They have a unique capacity to cultivate mushrooms and fungi. They use plants and wood that they bring to their nest through long tunnels to prepare spongy soil beds which are rich in cellulose or plant fiber. Certain types of mushrooms and fungi can grow on this soil. They are capable of breaking down lignin, a substance found in wood, into simpler organic matter.

The color of the termite mounds, called "tumuli," differs from that of the surrounding terrain. They are built from sand and clay dug from deep layers of soil. Whole colonies of mongooses live in these mounds and use air vents in the termites' nest as peepholes. The largest nests are those made by the goliath species, *Bellicositermes goliath*, of the macro-termite family. These nests are between 6 and 16 feet (2 to 5 m) high with a diameter of up to 65 feet (20 m). Pillars and pyramids and strange tumble-down castles are the work of the "warrior" termite and another species from southern Africa *(B. natalensis)*.

The pangolin, or long-tailed anteater, is a strange mammal. Its body is covered with horny scales, making it look like a reptile. It can roll into a ball like a pine cone. It has adapted so well to a diet based on termites and ants that it has no teeth. Instead, it has a long, thin tongue.

The aardvark is the only survivor of the order of *tubulidentates*. It has an incomplete set of teeth and an elongated snout with a tongue that slides in and out. Its saliva is sticky, so termites stick to its tongue. By night it moves from one termite's nest to the next. It never visits the same nests two nights in a row.

There is even a member of the hyena family, the aardwolf, which has become specialized and eats only termites. But this is the only case of its kind in the hyena group, which usually has a flexible diet. This miniature species of striped hyena has a small head. It can sneak inside the tunnels and corridors of termites' nests. Its teeth are small and not well developed, except for the canine teeth which it uses for fighting.

GRASSLANDS AND HERBIVORES

The grasslands of the savanna, as they appear today, evolved hand in hand with the ungulates, or hoofed mammals. The Graminaceae family of grasses has hundreds of edible species. The family typically has long, slender leaves and is the main resident of the grasslands. These plants have buds that permit growth to take place on the stem as well as on the top. For this reason, they are less vulnerable to grazing than other plants that have buds only on the top. To regrow the parts lost by grazing, these plants have developed increased photosynthetic capacity. Photosynthesis is the process by which plants use sunlight and simple elements to make complex substances essential for their growth. These plants also have the ability to shift their leaves so that shade areas are reduced.

Aside from rapid growth, grasses use other mechanisms to survive grazing. Some of them develop hardened tissues by storing hard silicon and tough lignin. Others develop coverings of spines or thorns. Some plants form highly poisonous or at least repellent chemical substances. Others grow together with species not favored by grazing animals.

The Effects of Grazing

There are small sections within African parks that look like miniature private plots. These are sampling areas, closed to grazing animals. They are set up to observe changes in the grassland community. Any changes observed here are compared to other areas where large herds of animals roam and graze.

Surprising results were obtained by studying a grassland where the main species were *Themeda* and *Pennisetum* grasses. Only four days after the migration of the wildebeest had begun, 85 percent of the grass had been eaten. But in the following month grass growth increased more in grazed areas than in protected areas. This is because seeds were being formed in the ungrazed area and growth was therefore slower. When the Thomson's gazelle arrived in the sampling areas, fences were removed. Tests indicated that gazelles ate roughly .035 ounces (1 gram) per square yard of grass in areas already grazed by the wildebeest. But they ate only one-third that amount in the sampling areas. The gazelles obviously preferred the shortened grasses of the grazed pastures.

If grazing is not excessive, it increases the growth of grass. During grazing, only older leaves are removed. They

Opposite: Two female impalas graze, choosing the most tender and nutritious parts of plants. They can be distinguished from the female Grant's gazelle, which is quite similar at first glance. The female impala has a black stripe along the tail and vertical bands on the thighs, which do not have white edges. An impala spends an average of 40 percent of its time feeding.

The gerenuk feeds solely on shoots and leaves, not grass. Its long, pointed muzzle is particularly well designed for tearing off small leaves from spiny acacia bushes. Its diet includes sixty-eight different species of plants in the Tsavo Park. It is probably even richer and more varied elsewhere.

contain a substance which slows down photosynthesis. Some species of grass, like *Andropogon greenwayi*, are dependent on grazing. They grow only in areas which are grazed, even heavily grazed. They develop a very tough, low-lying shape. Other taller and straighter species, like the foxtail grass, make up only 2 to 3 percent of total grass in grazed areas. In the enclosed areas, these same grasses make up 70 percent of the total. Being tall, they are better able than other species to compete for light. The balance between species in the struggle for survival is what creates the grassland/herbivore ecosystem.

Grass: An Ideal Food?

Digestion of grasses depends on symbiosis, the mutual interdependence between microbes and herbivores. Microbes live in the stomachs and intestines of herbivores. There they convert cellulose into sugar. Cellulose is an abundant plant fiber, but it cannot normally be digested by animals. The sugar that is converted by microbes is absorbed by the intestine.

In order to carry on this symbiotic relationship, some

Galls grow in reaction to the presence of ants. They form at the base of the spines of an acacia. This phenomenon is so common that a distinction is made between acacias which host ants and those which do not. This is an example of an animal-plant association. Some species of ants cannot nest outside these galls. The plant benefits from the association. The ants repel other insects and herbivores and keep other plants at a distance. The ants keep plants away by constantly removing everything that comes close to the acacia including seeds.

herbivores have a large sac in front of a series of three stomachs. This sac, called the "rumen," temporarily stores grass. During storage, grass is fermented by microbes. After grazing, the herbivores rest. As they rest, they bring grass back into their mouths and rechew it. This rechewing is called "cud chewing." Cud-chewing herbivores are called "ruminants." Ruminants of the African savanna include antelope, gazelles, wildebeests, and giraffes. It is not commonly known, but not all herbivores are ruminants. Some herbivores use another type of digestion. It is typical of rhinos, zebras, and elephants. These animals ferment grass inside a pouch in their intestines, not in the stomach.

Ruminant herbivores slowly and carefully digest their food. They also digest part of the microbes in the process. By contrast, large nonruminant herbivores eat rapidly and without pausing. They are better at grazing on tall grasses. Although tall grasses exist in sufficient quantity, they are high in fibers which are not nutritious. An elephant spends three-quarters of its day eating. If it also had to redigest the food in the rumen, it would have no time for sleeping. On the other hand, a waterbuck spends a third of its day grazing and another third chewing its cud. If grazing and cud-

Buffalo graze in a clearing with tall grass. These animals are not fussy about their diet. They eat a variety of plant species in various stages of growth. In the Serengeti, buffalo are found in most parts of the park, in areas with short grass. This is because, like other species of the Bovidae family, they eat by tearing at clumps of grass with their tongues. This tongue grip is different from the lip or tooth grip used by other species. So they have problems in grasslands that have been grazed too short. During the rainy season, they feed mostly on leaves. Leaves make up 70 percent of their total food. In the dry season, grass is their main food. Fifty percent of their total food is represented by stalks.

chewing hours are added together, the time spent ingesting and digesting food by ruminants and nonruminants does not differ much. Across time, herbivores have developed two different food-gathering and digestion patterns. Some use a short period for food-gathering followed by a lengthy period for digestion. Others take more time for food gathering and less time for digestion.

Fast regrowth seems to be nature's way of allowing grass to survive grazing. Regrowth is aided by large numbers of herbivores. They speed up the replacement of some plants with others. After digesting grasses, animals return nutrients to the soil in their droppings. This completes a nutrient cycle started with the absorption of nutrients by plant roots. Grass regrowth is stimulated in another way during grazing. The saliva of ruminants has a stimulating effect on grass growth. Regrowth is aided as herbivores also

Seasonal grazing occurs at different heights on a hill in the Serengeti. *Hill* is not the best term to describe the rolling lands of the Serengeti. Hills are actually low, long stretching upheavals in the terrain. But a few yards or meters up or down mean different water availability in the ground, and the growth of grass varies accordingly. Not much grass grows on the sandy and dry soil of the hilltops. Slightly more grows along bluffs and ridges. In the heavy clays of the valleys, grass is thickest and richest. Richard Bell describes the seasonal distribution of various African hoofed mammals according to their size and their feeding preferences along these vegetational belts. The small Thomson's gazelle (pictured at the top of the diagram and marked by a dot-dash line) prefer short grass. It allows them to spot predators more easily. They are the first to climb to the hilltops when the rainy season comes and the last to climb down as the dry season approaches. The buffalo (pictured beside and marked by a dash line) remain in the tall grasses of the valleys as along as possible. There they graze the grass short, preparing the ground for the zebra (pictured beside the solid line). Like the zebra, the wildebeest and the topi are medium-sized animals. Their feeding preferences are midway between those of the small gazelle and the large buffalo.

help in the dispersal of seeds. For several million years, grasses have evolved and survived. During this time a solid link has formed between grasslands and herbivores.

The Feeding Tactics of Herbivores

In spite of its vastness, the sea of grass is not an endless reserve of food. At the edge of the park, in fact, certain Masai areas are being overgrazed by domestic livestock. In a short time, overgrazing destroys the environment. This happens when the balance that existed for centuries is upset. The soil turns dry and cracks. It supports just a few scattered shrubs. The process of erosion interrupts the grass cover. The grass is short and sparse. The craggy outline of the hills is stripped bare by too many fires. Large trees are rare.

Savanna can easily degenerate into steppe. It could even turn into desert. Inside the parks, the threat seems remote and unlikely. For here the natural rhythms of life continue. Migration, rotating pastures, and a balanced use of resources exist, and the park supports an impressive population of hoofed animals, almost 2,500,000 head. This population includes a variety of thirty different species. Each species has different tastes in food and different environmental needs. This variety reduces the effect the animals have on grasslands and on certain types of vegetation.

The savanna offers a wide range of food options. There are many types of plants. And there are several vegetation

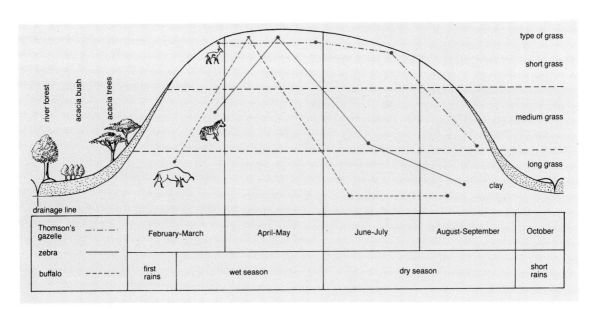

This diagram roughly shows preferences for the various layers of vegetation. The percentage of grass, leaves, and shoots consumed varies depending on the habitat and the season. The giraffe, at least in the Tsavo Park, can choose among sixty-six different species (mainly acacia trees). They share eleven of these with the gerenuk (standing on its hind legs in the drawing). There is much less overlap with the elephant, which feeds at similar heights, or with the black rhinoceros. The tiny dik-dik also has a liking for acacia, but it does not feed above 20 inches (50 cm). Thomson's gazelle feeds mainly on grasses at different growth stages. So does the kob (on the right in the drawing). The green guenon monkey (in the tree) is a gatherer of leaves, flowers, fruit, and seeds, as well as of tubers and roots.

levels. Plant types range from low-lying grasslands to long-growing grasses. Vegetation grows from grass to treetop levels. In addition, different habitats and seasonal variations help increase the number of environments. Vegetation provides hoofed animals with food, water, mineral salts, shade, and refuge. But it also provides a hiding place for predators. It is a barrier to vision, and it provides masses of inedible material to search through for favorite grasses. All of these factors relate to how an animal chooses a habitat. Size of the animal and its feeding habits are also important.

As a rule, smaller species require richer foods with higher levels of protein. They have a high rate of metabolism, and they can break down rich foods into energy and heat. The large herbivores are concerned with quantity. Through evolution this concern has led to medium-sized gazelles and antelope, including waterbuck or kob, and topi. The animals have narrow mouths, and flexible lips and

Following page: A large concentration of wildebeest and groups of zebra graze on tall grass during the rainy season on a Kenyan plain. A herd may number up to one million head and sometimes more. Both wildebeest and zebra feed almost exclusively on grasses of the Graminaceae family. The zebras eat a larger amount of the tougher parts (stalks) than the wildebeest. The wildebeest prefer the leaves. The buffalo and zebra are the first to move. The wildebeest follow. They prefer to graze where the older layer of grass has been removed and more nutritious parts are regrowing. The rate of regrowth of the grass during the dry season, in the northern regions, determines the death rate of the wildebeest. When there is not enough new grass, some of the wildebeest starve to death.

tongues. They have the ability to choose between leaves, stalks, and flowers. The large herbivores, like buffalo or wildebeests, have large mouths measuring 5 to 6 inches (12 to 15 cm).

The Impala, the Topi, and the Buffalo: A Case of Ecological Separation

The distribution of herbivores in the savanna is very logical. This is true in terms of the energy needs of the different species. It is also true in terms of their food and environmental preferences. One of the joys of a trip across the savanna is anticipating the possible presence of animals based on features of the environment. On such a trip, periods of waiting are followed by sudden discoveries.

One does not expect to see an impala when traveling through the extreme environments of open grassland or forest. This antelope is found in open wooded terrain dotted with the acacia trees. The impala needs shade, and it needs to drink if the water content in plants drops below 70 percent. The impala is a medium-sized antelope, weighing 88 to 100 pounds (40 to 45 kilograms). It is not migratory. But it moves about in a range of 250 to 2,500 acres (100 to 1,000 hectares). Its diet, consisting of 80 to 90 percent grasses, is selective. It carefully discards stalks and stems. These are the more fibrous and less nutritious parts of grasses. In the dry season, it will seek out acacia fruit, flowers, and leaves, but it avoids tall grasses. Its favorite tidbit is the small *Digitaria macroblephora* grass. This grass is green, tender, and has many leaves.

The same is true of the large antelope called the "topi." It weighs 220 to 310 pounds (100 to 140 kg). The topi has a narrow, elongated face. As it grazes, it chooses the leaves and sometimes the sheaths of grass, avoiding the stalks. It prefers open grasslands with grasses of medium height, 4 to 6 inches (10 to 15 cm). It likes *Digitaria* grass. It also likes *Pennisetum mezianum* in its regrowth stage after a fire. The topi's shape is easy to recognize. It has a slight humpback and lyrelike horns. It takes on a peculiar sentry position as it stands stiff and often upwind on hardly noticeable high ground in the grasslands.

The impala carefully selects both its habitat and its diet, but the buffalo can be found anywhere except in grasslands with low-growing grasses. It hardly distinguishes between plant species, leaves, stalks, or flowers. The buffalo has impressive bulk. It weighs from 900 to 1,550 pounds (400

to 700 kg), well over half a ton. It gathers a huge daily ration of forage, often of low quality. It shows a preference for areas covered with *Themeda triandra* and long green grasses. It prefers flowering specimens and grasses in an advanced growth stage. Only in the dry season does the buffalo become more selective.

The buffalo, the topi, and the impala are all non-migratory species. Their distribution in the Serengeti overlaps to a large extent. Competition and natural selection based on differing food requirements compose the foundation for ecological separation of these three species.

The Large Migratory Animals

The wildebeest, the zebra, and Thomson's gazelle make up more than half of the hoofed mammals in the Serengeti. Each year, they roam the wide grasslands looking for new pastures. Long columns of animals move from the southern plains, where they spend the rainy season, to the wooded lands of the north for the dry months.

Wildebeest herds on the march are at their most impressive. Trudging along in single file or several abreast, they move in a hunched gait as if they were tired. Then they break suddenly into a lope as they pour over hills and funnel down valleys, herd after herd, a living black flood tracing the age-old trails of their predecessors.

The routes of the wildebeest, zebra, and Thomson's gazelle overlap. As they move, they break hard tracks in the grasslands. But the tracks taken by the three species differ. Wildebeests move almost straight west. They do not reach the northern regions until August. Zebras, which take a more direct route, arrive in June. Gazelles leave the southern plains late and hardly ever reach the northern regions. Instead, they concentrate in a western corridor, between the Grumeti and Mbalageti rivers. Differences in routes and time for movement result from feeding preferences of the three species. These differences help the three species maintain ecological separation.

Zebras are not ruminants. They tolerate low quality grasses with a high percentage of stalks. They are the first to move toward tall grasses. They are followed by wildebeests. Wildebeests reach tall grass areas after the zebras have already cropped the grassy cover. The last to arrive are the gazelles. Tiny Thomson's gazelles demand top quality forage. They prefer to graze where the passage of the wildebeests has encouraged new grass to grow and has exposed

Right: The dik-dik is a miniature antelope. It has large ears and a more or less elongated snout, depending on the species. Its bent hind legs, make it look as if it were moving downhill. It has a tuft of hair on the forehead. In the male, this hides a small pair of horns. The dik-dik is a typical inhabitant of savanna and bush, and arid areas with a good cover of acacia. Acacia trees form a vital part of its diet, which is based on leaves, shoots, and fruit, as well as roots and tubers.

Below: The map shows the migratory routes of the wildebeest in the Masai Mara-Serengeti, between 1969 and 1973. After the rains in April-May, they move northward for the dry season. After the short rains in October, they move southward again.

WILDEBEEST MIGRATIONS (1969-1973) IN THE MASAI MARA-SERENGETI SYSTEM

- December-April
- May-July
- August-November

low-lying plants. These are usually hidden by tall grasses which are very rich in protein. Grant's gazelles are more than twice the size of Thomson's gazelles. But they are even more selective in diet. They feed on plants which absorb moisture at night. As a result, they can go for some time without drinking. In addition, they can stay in the arid southern grasslands for much of the year.

Migratory hoofed mammals live in the lowlands only when there is enough rain for the grass to grow. The areas they visit undergo irregular periods of heavy grazing depending on herd movements. At times, herd movements help reduce competition with nonmigrating species. In general, the waves of migratory animals do not greatly disturb nonmigratory species. The impala changes its diet by including acacia shoots and pods during the dry season when the wooded regions are invaded by migrating animals. During this period, buffalo move to the dense, green belts along rivers. These areas are usually ignored by nomadic herds except for access to water. As a result, the habitats of different hoofed mammals hardly overlap at all.

Competition creates and stabilizes the division of

This diagram shows feeding levels of the elephant and percentages of food eaten. The elephants' diet varies with the seasons. The percentage of grasses consumed increases in the wet season. In the dry season, it takes more food from trees (leaves, branches, and bark). Feeding levels change depending on whether its habitat is open grassland, bush, or forest growing along streams. Even if plenty of grass is available, the elephant will not restrict itself to just grazing. It rounds off its diet with shoots and leaves which are often more nutritious, and with bark. Bark is digested swiftly because of its high fiber content. It is rich in sodium and calcium. These minerals are necessary for the continuous growth of the tusks.

resources among the herbivores. In certain cases, the grazing cycle of larger species actually prepares the ground for smaller species. Energy spent during the long migratory march is repaid by continual access to new grazing grounds. These grounds can withstand instant and intense exploitation. Migrating animals have an advantage over resident animals because they can completely consume resources in an area and then move on.

It is the "black tide" of wildebeest which dominates the herbivore community. Their feeding strategy involves grazing over wide areas. It is an excellent adaption to the savanna and to its distribution of resources. The distribution is scattered, cyclical, and varying in terms of climate and fires.

The Elephant Problem

During most of the year, the Serengeti ecosystem includes two distinct elephant populations. One lives in the north, the other in the south. Their population density varies between one and two head every 4 sq. miles (10 sq. km).

During the last twenty years, there has been a marked increase in the elephant population. This increase is due

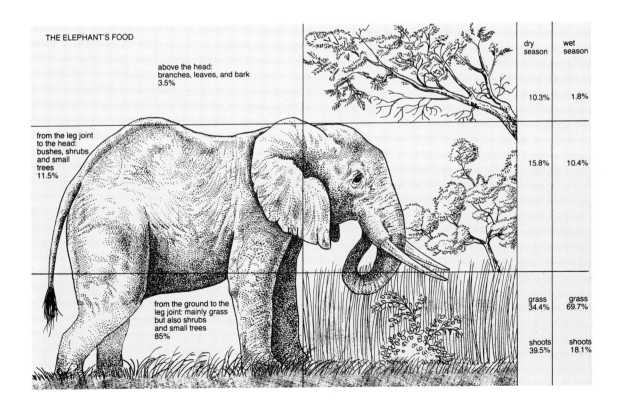

39

An elephant tries to reach the green foliage of a tree with its trunk. Sometimes, for various reasons, it may even knock the whole tree over. It is also possible to see elephants making unexpectedly nimble jumps to reach higher branches. The passage of elephants is marked by uprooted trees or trees without any bark for the first 6 to 10 feet (2 to 3 m). Sometimes, however, they are accused of destruction for which they are not to blame. The death of many trees in the Amboseli Park in Kenya, which seems to be the work of elephants at first glance, is in fact due to an increase in the salt content of the soil. Strong gusts of wind can also make a landscape look as if elephants have just passed through it.

mainly to immigration through the boundaries of the parks. The immigration is caused by the spread of cultivated lands surrounding the parks. The presence of elephants does not go unnoticed. Elephants graze during the wet season in regions of open savanna. They also feed on bark ripped from tree trunks, branches, and foliage. During the dry season, trees become the main ingredient of their diet.

Trees damaged by elephants inevitably die. Elephants often uproot trees while getting at foliage or while scratching on them. Dominant males may uproot trees to show off their strength. Later, such trees are burned or devoured by termites. However, not all of the elephants in a herd are experts in uprooting trees. Some try it over and over again and do not succeed; others never try. Some bulls get very adept at pushing over trees. They make the tree swing by pushing with the base of the trunk from different sides and then choose one side for the final push with the lower or the upper base of the trunk.

Several models propose ways to explain fluctuations in the numbers of elephants in the pack. They illustrate cycles of population density. One model suggests a sudden drop in population every fifty years due to drought. The drought is followed by a period of thirty to thirty-five years during which trees grow back. This model holds true if no other destructive events occur in the area during the time. Another model does not include a climatic cycle. It is based on the simple density association between trees and elephants. This cycle repeats every two hundred years. In this mode, an increase in vegetation favors elephants. When they become too numerous, they cause a reduction in the number of trees. Tree reduction slows the rate of increase of elephants, and the cycle starts all over again.

Both of these models suggest that it is unwise to interfere with natural rhythms. They suggest that rhythms come into balance naturally. They suggest the possible need for a slaughtering plan for elephant population control. Recent studies indicate that before the arrival of people, elephants roamed over extended regions. Periodically, they migrated in search of food. A herd returned to the same area every one to two hundred years.

Today, elephants lack migratory possibilities. The heavy demands they make on resources explain the cyclical population pattern which exists in park areas. It is not realistic to expect the return of a natural balance in populations caught within artificial boundaries.

HERBIVORE SOCIETY

Opposite: A group of kongoni inside the Ngorongoro Crater shows their distinctive "downward" shape and their long, narrow heads. Their horns grow from bony projections at the front of their heads. Their habitat is the open savanna as well as the bush, as long as it is not too dense and arid. The kongoni form groups of four to fifteen individuals. At times, their groups number between thirty and fifty head. Assemblies of thousands of head, described in past days, are very rare. The group includes one male with several females and young. The group remains stable throughout the year. The old males lead a solitary life, while the younger bachelors gather in small nomadic groups. These are the most settled of all the large antelope. Each group has a territory. It is marked out with scents secreted by the male, as well as by the females and the young. The kongoni only migrate if there is a shortage of grazing or water.

The world of the savanna is teeming with life. It provides both a close-up and an overall view of savanna life forms. Hordes of wildebeest draw attention from less obvious, hidden forms of animal life. The relationship between diet and behavior was studied in seventy-four species of African hoofed mammals. The studies looked at the territorial, reproductive, and antipredator behaviors of these animals. It was learned that as the body weight of an animal increased, its diet became less selective and it lived in a larger social group. High protein plants are scarce, scattered, and available only at certain periods. The search for such plants is almost always carried out by individual animals. However, a group of animals can best exploit wide stretches of grassland. As animals group together, the social structure becomes more complicated.

How Food Availability Affects the Social Structure: The Case of the Antelope

For a pair of dik diks, a miniature antelope weighing only about 9 pounds (4 kg), a kopje may be the center of the universe. Schaller writes that two animals sometimes stood just outside a shrub and watched him pass, the tuft of hair on their heads erect with excitement. As soon as he stopped, they vanished. Only their shrill nasal whistles revealed the direction they took in the dense undergrowth.

A dik-dik stares at a potential predator with wide eyes. It stares not out of curiosity but out of fear. At first, fear paralyzes the dik-dik. Then it flees to cover, sending out alarm whistles as it runs. This whistling makes up for a lack of visual signals in the thick of the bush. The penetrating sound it makes, "zik zik", gives the dik-dik its name.

Tiny and vulnerable dik-diks camouflage themselves in the shade of acacia trees and bushes. They have a keen awareness of their territory, which covers only a few acres. They defend this territory against members of the same species. It is marked out at precise spots with excrement, urine, and secretions from the orbital glands situated near the eye sockets. Dik-diks use mainly their sense of smell to communicate with each other.

Many small antelope are highly selective in their diet. Their diet is based on shoots, flowers, and fruit. They lead a solitary life or live in pairs. They stay in the same territory their entire lives. This territory must be large enough to insure a year-round source of high protein food, but small enough to be defended.

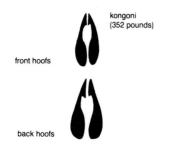

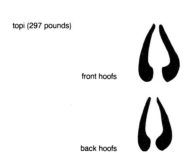

Above: The hoofprints, front and back, are of the wildebeest, kongoni, and topi, the most widespread of the large antelope (subfamily: *Alcelaphinae*).

Slightly larger antelope, like the gerenuk, also live in a territory. There they are completely familiar with plant species and the best time to eat them. The gerenuk grazes on the tender leaves of bushes and small trees. It chooses food that is preferred by the giraffe, which it vaguely resembles. In fact, in Somali, *gerenuk* means "having a giraffe's neck." The gerenuk stands up on its hind legs. It rests its front legs on the tree trunk, and extends its neck. In this way, it is able to reach up into the highest branches. This rather unnatural posture could be compared to that of the praying mantis. Like the oribi and other medium-sized antelope, the gerenuk lives in pairs or in small groups of five to ten individuals. It spends all year in the same area.

Larger antelope often gather in large grazing groups. The kob, an antelope with a long, coarse coat, enjoys this type of social pattern. Males spend the first months of their lives in the females' nursery. Then they gather in groups of five or six. The young do not establish an individual territory until they are six or seven years old. Such a territory varies in size, ranging from a few acres to more than a couple of hundred acres. Size depends on the distribution of resources.

The female's story is quite different. When female kobs leave their mothers, they lead a nomadic life for a few years. This helps them avoid inbreeding, mating between closely related animals. Eventually, they join a herd of other females. Bonds between females of the herd are not as strong as bonds between males in herds. The female herd passes through the territories of several males. There they

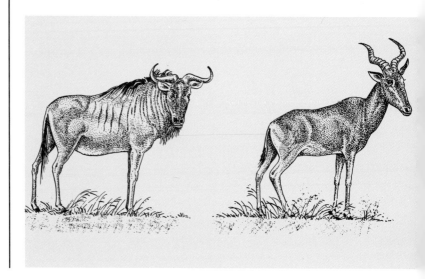

Antelope belonging to the subfamilies *Alcelaphinae* commonly called "deer antelope," and *Antilopinae*, commonly called "gazelles" are drawn to scale. *From the left:* The wildebeest stands 52 inches (132 cm) tall at the shoulder. It has a heavy head and a black or white beard, depending on the subspecies. Lightly colored forms occur from central Tanzania to the Tana River, in Kenya. Darker colored varieties live in more southerly areas. The kongoni is 48 to 52 inches (122 to 132 cm) tall at the shoulder depending on the subspecies. Coke's kongoni is lighter with a sandy coloring. Jackson's kongoni is larger and reddish brown. Grant's gazelle stands 31 to 35 inches (80 to 90 cm), tall at the shoulder. It has a black stripe on the hind edge of the thigh. The impala stands 36 to 42 inches (92 to 107 cm) tall at the shoulder. It has lyre-shaped horns and a more brownish yellow coat as compared with Grant's gazelle. Its vertical black thigh stripe has no white-edges. It has two black markings on the hind legs. Thomson's gazelle stands 25 to 27 inches (64 to 69 cm) tall at the shoulder. It has a distinctive black crosswise band on the flanks. The topi stands 48 to 50 inches (122 to 127 cm) tall at the shoulder. It is a brownish yellow color and has black markings on its legs.

remain for a few days or a few months, without forming stable pairs.

To the observer, such temporary gatherings may seem like a harem. This is true of the impala. One single male may appear with his long, lyrelike horns in the middle of a large herd. The herd may include twenty, fifty, or even one hundred females. The male impala tries to keep control of the herd. He makes sure that it stays within the borders of his territory. He also chases off any rivals. He threatens them or engages in battle. Differences in the physical appearance between sexes increase when the male is engaged in intense competition for territory females.

Groups of female impala move from one male territory to another. This is because the territory of a single male cannot provide enough food throughout all seasons. Impalas' food needs cause changes in their mosaic-like territories. A male's territory changes according to his females' habitat preferences. Males limit their areas in the rainy season, when grass is abundant. During mating season, territories come closer and closer together. During dry periods, when there is not much grass, these territories may disappear altogether. Often, mixed groups numbering more than one hundred individuals form.

The most impressive gatherings of individuals occur among the larger antelope. They feed solely on grass, which is a widespread and plentiful resource. These animals do not need to know their grazing area well. Antelope like the wildebeests have a diet that depends only on the stage of growth of the grass. They form endless lines numbering

A group of impala gathers at a watering hole. The photo shows the harem-type social structure. Females and young gather around a single male. The male is larger with lyre-shaped horns and stands in the middle. He is probably on the alert because of the presence of the photographer. The male maintains a territory rather than a precise harem. Control of this area extends to the females entering it. The average size of the group is about fifteen to twenty individuals. But during the dry season, larger groups are formed. The impala need to live close to water. But they can survive on just dew for long periods.

thousands of head. The wildebeest is a strangely-shaped antelope. Its head is heavy, it has a shaggy white beard, and its curving horns make it look angry.

The mating season coincides with a nomadic phase of life. During this season, males establish small temporary territories. Territories last only a limited period. They do not affect the structure or purpose of the "super-herd." The herd intensely uses available resources, grasslands covered with new fodder. This use is without any organization based on age, sex, or family.

The largest of all antelope are the eland and the oryx. The eland has a distinctive striped back. The oryx has long black horns. Buffalo belong to the Bovidae family, which is the same family as that of antelope and gazelles. Some species of this family feed both on grass and on leaves and fruit. They are not very selective in their choices. These giants of the grasslands gather in huge herds. They gallop

about like thunder, males and females together with the young by their sides. The social order that determines mate selection does not lead to the establishment of territories or harems. This would split the unity of the group. A herd of buffalo may reach a total of two thousand head, but it still retains a stable identity and composition. Leading it will be an elderly female or cow. Large males or bulls stay near the rear. Should a predator attack, all of the herd members cooperate to repel it.

The Group as a Defense

Group formation has benefits as a feeding strategy. It permits an intense and complete exploitation of grazing grounds. It is also an important factor in defeating predators. In many cases the risk of being hunted is lessened by strength in numbers. In a mass of animals, a predator has less room to move about and is hampered in its attack. Also, it is possible for the chosen victim to escape by mingling in the crowd. This is called the "diluting effect" of the herd, and it is common among the antelope.

An attack is always aimed at a specific animal. It is never aimed at the entire group. When a predator charges, the herd bursts out in all directions. This leaves the enemy confused and disoriented. It wastes time homing in again on its victim. It may even stop in its tracks to consider its choices. For the antelope, the technique of bursting out works well against lions. It works because lions are not very disciplined in their attacks. It does not work as well against cheetahs. The cheetah remains focused on prey it has singled out. Against this enemy, the main defense is swiftly fleeing, possibly with sudden swerves. A gazelle, for example, runs at speeds of 56 miles (90 km hour) in a zigzag pattern. Impalas leap acrobatically into the air as they flee. Wildebeests make diagonal jumps.

The way wildebeests move, in narrow columns, is also an antipredator strategy. They leave behind a scent, a secretion from glands between the toes, beneath the hoof. The scent is so strong in the grass that even a human's weak sense of smell can detect it. Following this scent, the wildebeests move in a single track. They move along in single file rather than in a fan-shaped marching formation. This reduces the risk of stumbling upon a hidden predator.

Almost all animals, including the usually silent giraffe, can whistle, roar, neigh, bellow, or snort when they see an enemy. Thus they can signal danger to other members of a

A female Thomson's gazelle is "stotting." This is a jump made with the legs held stiff. It is typical when danger is present. The picture on the left shows the usual "stotting," with the gazelle just a few inches off the ground. The middle shows panicked, very high "stotting." On the right, the gazelle lands.

group. The Thomson's gazelle and many other antelope give a visual alarm called "stotting." The gazelle makes an exaggerated leap upward with legs held stiff and the tail extended. Then it bounces back on the ground.

By sounding an alarm, an individual makes itself apparent. The animal loses the advantage of being lost in the crowd. The antelope who sounds the alarm puts the group on alert. By jumping, the animal also indicates its physical ability. Such athletic display may discourage the predator from pursuing it. In addition, the antelope uses the

A family of warthogs travels under the watchful eye of the older male (on the right). Warthogs are 30 inches (76 cm) high at the shoulder and have a brownish gray hide. The warthog can often be spotted in tall grass because of its tail which is held upright when it moves. It has a short neck so when grazing, it bends its front legs in an odd "kneeling" position. It lives in open savanna, always close to a watering hole, and rests during the hottest hours of the day. As a rule, families live with young belonging to one or two successive litters. Older males live on their own. Several groups may temporarily join together. But each family group retains its own identity.

Following page: Burchell's zebras graze in the Masai Mara Park. In the foreground, the photo shows a flock of Egyptian geese with brown markings around the eyes. In the background, Thomson's gazelle and an impala are grazing. From a distance, in poor light, the zebra's coat looks gray. As a rule, their stripes stand out clearly against the backdrop. It is easy to see zebras. But is it as easy to pick out just one zebra? One theory explaining their strange coloring suggests that the mimetic effect of zebra in groups confuses predators.

Below: Two zebras stand in the typical cross-over position. This is a friendly act which is probably used for shared sentry duty. It gives an all around view of 360 degrees. In the Serengeti Park the zebras are prey to lions, especially in the period when the young are born. At this time, they concentrate in the southern plains.

herd as a distraction. It sets the herd to flight. The animal who spots an enemy has the advantage of being the first to see it. It is the only animal in the herd that really knows which way to run as the herd takes flight. On some occasions, a close bond of kinship triggers altruistic behavior, acts which help others. This may involve protecting young or other closely related animals.

Living in a herd also means having one's senses multiplied. The antelope's eyes jut out to achieve a wide-angle vision. But as part of a herd, the power of its eyes, ears, and nose is increased. Its chances of detecting the presence of an enemy are increased. In a herd, it is likely that at least one individual is always on guard. The territorial and solitary male pays a high price for the reproductive advantage of having a territory. Other acts also work to make the group a defense mechanism. The topi and the kongoni often stand upright on high ground or on a termite's nest. This greatly reduces the risk of being preyed upon. The same is true of the friendly stance of two zebras. One of them rests its head on another's back. Together, they have a 360-degree view. An alarm may also travel between different species. When an alarm spreads, it causes a simultaneous flight of topi, impala, and kongoni. One moment, they may be peacefully gathered around the same watering hole. The next, they may be in flight.

The warthog is a species similar to the wild boar. It lives in families of eight to ten individuals. Parents live with one or two litters. It often mingles with antelope because its short legs do not allow it to see very far. The warthog is one of the lion's favorite prey. Sometimes it grazes among buffalo. It uses the protection offered by their massive ranks. It has a short, stiff mane and moves with its tail held upright like a flagpole. It moves about with short, restless steps.

The large male weighs up to 265 pounds (120 kg). It takes up the rear, while the female moves off with the young. Warthogs are frequent residents of the savanna. They are important because they have an ecological niche, a place in the environment. They have an unusual and complex relationship with various species. Antelope act as their sentries. Buffalo are their protectors. Abandoned lairs of aardvarks are their refuges. Their favorite pasturelands are patches of low, trampled grass left by hippopotamuses. From there, warthogs have good visibility and can spot enemies.

When raising their young, herbivore mothers defend them to the end. They are very effective, especially if they

have horns. In addition, they use an indirect but effective group strategy. Females all give birth to their young at about the same time. The young of the wildebeest are all born within a span of three weeks, during January and February. The zebra also follow this pattern, but the birth period is longer. Predators can only eat a certain number of animals. Because many young are born at the same time, their overall chances of survival increase. If births were more spread out, fewer young would survive.

Immediately after birth, a young wildebeest can stand up on its own. At first it stands with difficulty, its four legs spread wide. In ten minutes it manages to walk, staying close to its mother. The loss of calves is a tragedy which recurs every year in the wildebeest's world. Every time the herd takes to its heels, many young calves run astray. They wander, searching desperately for their mothers through the moving herd. Often they have little hope of success. A lost or abandoned calf is hardly ever adopted and almost certainly dies of hunger unless it is eaten first by an enemy.

Generally, when one member dies, the herd is indifferent. Immediately after a hunt, the herd reforms and starts grazing again. Though the group may seem indifferent, it still represents a source of safety. Various reports from the Serengeti Park tell stories of old buffalo seeking new sources of safety. When isolated from the herd, they head toward a ranger station to die. When their strength fails them, these old animals feel safer close to human beings than alone in the bush. In the bush, they might be torn to bits by a hyena or a lion. They prefer to die a natural death.

The Harem, the Family, and Solitary Life: Other Herbivore Behavior

Antelope are the most numerous animals in the Serengeti ecosystem. Across the grasslands, zebra often form long, striped columns. They gather in scattered herds or mingle with wildebeest, their black-and-white bodies dazzling in the intense light. Even when they assemble by the thousands, it is possible to make out small groups of bachelor males. They are isolated from family groupings. Family groupings include a stallion, half a dozen females whose positions are based on their rank, and foals beside their mothers. Family groups are based on long-term associations between an adult male and a group of females. The male does not establish a territory but exerts direct control on his females.

A dominant male keeps other males well away from his mares. He competes with them in his courtship of an adolescent female in heat, ready to mate. In order to start a harem, a male must either defeat another male in a fight or kidnap young females. Males voluntarily leave their parents around the age of two. Just before they turn two, the female young are "kidnapped" from their families by males from other families. When she is two years old, the female becomes a stable member of a family.

Within the family unit, the stallion is closer to its young than to its females. It is perhaps for this reason that defense rests with the stallion. He defends the herd from lions, hyenas, and hunting dogs. Only rarely are mares in the group involved with defense. When a herd of zebras is on the run, the stallions defend the rear of the herd. With their ears lowered and their teeth bared, they stop only to bite their pursuers, filling the air with loud whinnies.

The defense put forth by a group may be passive. One example is the protective wall set up around individuals under attack. An active defense puts forth an aggressive response to predators. Stallions may bite and kick. A buffalo may charge. An elephant's attack may terrify. Its pillarlike

Above, on left: A diagram from above shows how a long line of zebras moves. The pointed end of the shapes indicates the head. On the right, the diagram shows how zebras scatter into individual family groups following a disturbance. The scattered pattern takes place as soon as one individual in the line slows and stops.

Right: A family group of elephants takes defense formation. The adult females stand in the foreground and the young behind them. Danger or the show of fair hostility is indicated by the upright body position, with the ears spread forward and the trunk extended. The tree on the left is a broken acacia which has been stripped of some of its bark by the elephants.

Opposite, top: A diagram of the defense formation of a group of elephants is seen from above. The group includes females and young elephants. If the group retreats, the dominant female defends the rear. She turns to face the enemy and threatens to charge. If the group takes to its heels, its speed matches the pace of the slowest young animal.

legs are the size of tree trunks. They can form a barrier that is not easy to get past. This provides excellent protection for the young. In a group of elephants, it is always the dominant female who takes the initiative. She may decide to attack, to move out, or to close ranks. Elephants form matriarchal, or mother-dominated, families. Their families include an aging female, two or three daughters, and several young of varying ages and sizes. As the females grow, they stay with the family. But at puberty, the male young go off and live alone or with other males. They form loose and temporary groups which periodically visit the family. At the start of the rainy season, several families may join together to form groups of about fifty individuals. But each group retains its own identity.

When an elephant is killed, the matriarchal female abandons it on the ground only after a long period of time. When an old female dies, the entire family stays near her. This habit is well known to ivory hunters. It makes it hard for them to get their kills. When elephants are on the move, the senior female is at the head. She is the one who chooses the grazing grounds. Her reproductive capacity declines with age. As she ages, she uses her lifetime of experience in

A mother black rhinoceros grazes with her young inside the Ngorongoro Crater. A few oxpeckers on their backs search for parasites. Black rhinos are usually found on their own, or in mother-young pairs. Large mounds of droppings are used as social signals to maintain distance between individuals. But rhinoceroses do not mark out an exclusive territory. Adult males, in the Ngorongoro Crater and the Olduvai Gorge, spend about 80 percent of their time alone. Crossbreeding between related individuals is avoided because a mother will not put up with her larger male offspring when there is a newborn. She will chase him away, even ot he is still weak compared to his enemies. As a rule, young animals gather together or seek out an adult female. As they wander about, they leave their birthplace behind.

service to the family. She lives an average of 60 to 70 years. But some live to the ripe old age of 120. It is the matriarchal figure who remembers events which occurred in the past and who is probably the basis for the legend of the amazing memory of the elephant.

White rhinoceroses live in families of four to five members. They have their own grazing areas and fixed drinking holes. They are not a dangerous species. They are placid by nature and only rarely charge. No family structure is found with the black rhinoceros. Males are solitary. Bonds between females and their young last only as long as the young require protection.

The reputation for being unsociable and ferocious that clings to rhinos is mostly legend. Their individual territories may overlap slightly. This leads to frequent encounters between two rhinoceroses. The meetings include proper

greeting rituals. They approach cautiously and touch noses. This is followed by a series of gentle taps with the head. There is no clash between residents who know one another, but a stranger that happens to stray into their territory will be violently attacked.

The tendency to charge at humans, as shown by rhinos in some areas, may be caused somehow by traditional hunting methods. A rhino in flight may easily be hit by a poisoned arrow. But a more aggressive rhino who charges may escape such a fate. Thus, more ferocious offspring have an advantage against traditional hunting methods, using bows and arrows. On the other hand, the use of guns in some areas tends to favor more timid individuals because a charging rhino is more likely to be felled. The only escape from a gun is flight. In areas where the people are herdsmen, rhinos have never been hunted. There, rhinos are the most docile creatures in the world.

The Mating Season

In the mating season, males of the same species often engage in seemingly violent battles. But usually no blood is spilled. Such fights take the form of a ritual. They have definite and repeated patterns and movements. They establish the dominance of one male over a territory and/or a female while keeping damage and loss to a minimum.

It is quite difficult to determine just how violent clashes between male giraffes really are. All of the movements of this species look as if they were filmed in slow motion and without a sound track. They do not seem violent at all.

Giraffes form unstable and scattered groups, with young and adults of both sexes mixed together. They do not develop any mutual defense strategy. Sometimes males band together, exhibiting some degree of dominance, or they may live a solitary existence.

Giraffes have a very selective diet. They eat treetop foliage and fruit, so they tend to eat on their own. But in open plains strewn with acacia trees, contact is maintained among giraffes through a system of visual communication. This system appears to work even when individuals are far apart. While grazing on vegetation, giraffes shape trees in a typical way. It is as if they are careful gardeners. The tallest trees are shaped like cones, and young trees are kept the size of bushes, about three feet (1 m) in height.

When a female is in heat, many males are attracted to her. They engage in battles which are ritualized shows of strength and power. Such rituals and other forms of communication within the species often revolve around the most-developed physical features. Thus, elephants confront one another by intertwining their trunks. Giraffes sway their very long necks, standing beside each other with their stiff legs spread out. Hammer-blows struck with the head and the neck are rare, but can be heard from afar. They are powerful, too. A giraffe has been known to send a half ton eland flying into the air with a single blow. But neck damage is minimal because the hide on the giraffe's head and neck is thick. It forms a sort of shield. This protects the giraffe from powerful blows that might otherwise be lethal. It reduces the risk in battle without lessening the effects of the signals.

Wildebeests confront one another by shaking their heads and stamping on the ground with their hoofs. Sometimes they drop to their knees and rub their horns in the

Male giraffes face each other for a "neck duel." The two contenders stand side by side. Suddenly one of them lowers his head and sweeps it against the other's body. There is much huffing and puffing which can be heard some way off. These clashes probably establish dominance. They occur in groups of bachelor males or between isolated males who meet. Among isolated males, the issue might be the defense of a territory. But there is still much debate regarding territorial behavior among the giraffe. In the wild, groups and isolated individuals are equally common. It is probable that some duels relate to reproductive behavior. They occur all year round. There is no precise mating season for giraffes. The young are born at different times of the year, after a gestation period of 450 days.

Different types of duels occur between different male animals. The drawing on the left shows how two giraffes clash. They swing their necks like maces. At the top right, two kobs, or waterbuck, clash. They confront each other with blows of their horns. In Aberdare Park, Kenya, a duel between two males ended with the death of one. But this is quite rare. At the lower right, two zebra stallions bite and kick each other.

dust. They get up and bend low again. Then they clash together with their horns, as if by some agreed signal. They move forward on their knees until one surrenders, jumps to its feet, and runs off. They try to twist the opponent's neck and knock him off balance, but rarely do they give or get more than a slight scrape on the head or neck. The victor chases the other with a peculiar dance.

The territory established by a male wildebeest may be kept for a few days or just a few hours. He stays in the same small area only long enough to mate with females during a migratory run. The term *pseudo-territory* is used to describe the area. These small areas are sometimes reduced to circles measuring only 36 feet (11 m) or so across. Here the male gallops about with an odd and unmistakable gait. He tries frantically to surround and isolate small groups of females in heat.

The degree of violence during fights is related to a species' territorial behavior and reproductive strategy. The clashing of horns between two impala can sound very loud. Impalas fight for control of a whole group of females waiting close by. The air is filled with astounding bellowing.

The drawing shows more examples of duels between males. At the top left is a duel between impalas. Their duels are limited to the mating season and are accompanied by noisy grunts which tell of the animal's presence in a territory. Below, the drawing shows how two kongoni fight. They confront each other by shoving their bulky bodies together. Their heads are protected by the thick, bony growth on the forehead. On the lower right, two wildebeest are shown in the typical kneeling position. This position is also used by the kongoni.

This is part of both duels and mating. It attracts new females and new contenders. The cost of holding onto a territory is high, but it is well repaid by gains in reproduction.

The kongoni are long-legged antelope with a strange, swaying horselike gallop. They live within a permanent territory, like the impala. They live where enough grazing and water are available.

Kongoni males establish territories. They keep watch over their group of females and young from a choice location. They often stand upright on a termites' nest, for example. Bachelor males are not permitted in the herd's territory. They form nomadic groups. In their duels, males suffer serious injury or even die. Such duels are more frequent and longer-lasting when females are ready to mate. But the case of the kongoni is an exception. Among antelope and gazelles, only ritual aggressive behavior is the rule. These are among the more vulnerable herbivores.

Thomson's gazelles challenge one another with clashes of horns. Their duels are short, elegant, and so contained that they look like games. But many a male ends up with his horns broken or blunted.

CARNIVORES

Humans have always had a certain admiration and fondness for the large carnivores. Perhaps they identify with them in their role as predator. Perhaps they feel a relationship to animals which have developed a high level of intelligence.

Problems of hunting require the development of intelligence in carnivores. These animals understand complex situations. They have an ability to foresee possible developments in a situation, and they are able to make decisions quickly. These abilities are aided by an impressive learning ability.

In certain carnivores, the advantages of group hunting have triggered the development of social structures. These structures are also associated with a high degree of intelligence. Flesh is a very nutritious food, and it comes in a concentrated form when large prey is hunted. Herbivores spend most of their time eating. Almost all the carnivores, on the other hand, suffer from having too much spare time. They have found a solution to this problem; they spend most of their time sleeping or resting. Some species, however, dedicate their leftover time to social relations.

Hyenas, lions, hunting dogs, and jackals establish within their groups precise dominance levels and social roles. Each animal also shows temperament, individuality, friendship, hostility, and rivalry. In short, animals have qualities used to describe humans and to some extent monkeys and apes. Such findings are the result of careful and rigorous research. They are not just a sentimental view of the animal world.

The Predator-Prey System

Competition between and within species is the dominant factor that results in the evolution of a species. Over time, a stable and balanced species develops. The predator-prey relationship is similar, in many ways, to the relationship between plants and herbivores. The action of herbivores benefits vegetation of the savanna. The action of predators also has a beneficial effect on the animals they prey upon. This is true in spite of the fact that the immediate needs of the two groups seem to be opposite.

Predators and prey evolve together. They live in a state of balance which tends to maintain itself. An increase or decrease in prey leads to an increase or decrease in predators. Across time, balance is restored. The predator-prey interaction can also vary depending on other factors. A

Opposite: The male lion, as recent studies have shown, is a lazy hunter. Whenever he can, he feeds on the kills made by the females in his pride. Solitary males do not have a territory. They have no females and are forced to take care of themselves, But like all lions, when possible, they steal prey killed by others by asserting their physical strength.

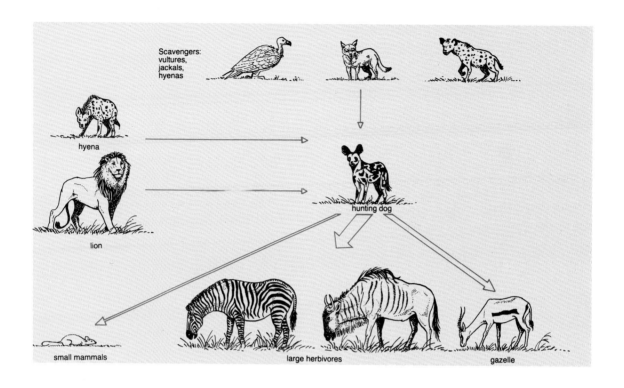

The hunting dog, now rare, occupies a central position in the ecosystem of the savanna. It can kill large prey, leaving tidbits for the scavengers, vultures, jackals, and hyenas. But it is often chased away from its own kill by groups of hyenas or lions. The current rarity of the hunting dog may be explained by the increase of hyenas in protected areas. In the diagram, different sizes of arrows point at the small mammal, the zebra, the wildebeest, and the gazelle. The sizes of the arrows are in proportion to the importance of such prey as food.

major external factor that influences the quantity of prey is fodder availability. This varies according to the seasons. Observations suggest that it may be unwise for humans to interfere with the natural predator-prey system by artificially reducing the number of predators.

Nevertheless, in the 1950s, the hunting dog was actively hunted, even in conservation areas. In the hope of increasing herbivore populations, the hunting dog was exterminated in Ruwenzori Park in Uganda. The desired effects did not come about. Today, people are very aware of the role played by predators. Efforts are being made to reintroduce the African hunting dog in the park and in other areas where it has been eliminated. Unfortunately, the population of this species has dropped dramatically in all areas where it has been hunted.

The elimination of all predators triggers an explosive increase in prey population. This, in turn, causes a rapid deterioration in the environment due to overpopulation. Ultimately, this approach harms the species it was intended to help. The carnivores have developed natural systems of self-regulation for their population density. Using their system, overuse of resources is avoided. Predators play a beneficial role in the natural selection of the species they prey

A spotted hyena feeds on the entrails of a Grant's gazelle in the midst of an attentive flock of griffon vultures. Hyenas are able group hunters, but they are also scavengers. In this role, they adopt a plan of individual food-searching. This way, they are more likely to find carcasses. The animal in this photo was probably guided by a flight of vultures. It found the gazelle recently killed and left almost intact. But it is still very much on the alert, the rightful owners of the prey are probably still close by. They may come back to assert their rights.

upon. They kill genetically weaker and less fortunate individuals. In doing so, they remove those from the breeding population. Predators also prey upon ailing animals, thereby improving a prey population.

Hunters or Scavengers?

The carnivores of the savanna are traditionally divided into two groups, depending on their feeding habits. One group includes the scavengers, such as the jackal and the hyena. The other contains the noble predators, such as the lion. In truth, a group of lions feeding on prey at dawn may have stolen their prey from hyenas which killed it during the night. The lions may be surrounded by a group of disgruntled hyenas waiting for a few tidbits.

Studies prove that the distinction between scavengers and the so-called noble predators is valid. The spotted hyena, considered a scavenger par excellence, is actually an active predator. In the Serengeti, it hunts more than 60 percent of its own food. The lion, on the other hand, will feed on animals killed by others whenever it can. There is

A cheetah has just caught a gazelle. It is still alive and probably immobilized by an injury to its backbone. In this type of situation, the predator does not kill the victim by suffocating it in the normal way. (See the drawing on page 79.)

often competition between hyenas and lions over slain prey.

Sometimes, if they are numerous enough, hyenas push away a young lion or a lioness. But they can do nothing about an adult male. In these clashes, the male lion is very aggressive. It will not just chase hyenas away. It will kill them if it can.

Not all of the duels between carnivores are bloody. Many are resolved with a swift trial of mutual strength. Serengeti hunting dogs weigh about 55 pounds (25 kg). They lose about 50 percent of their prey to lions and to

hyenas, which weigh almost twice as much as the dogs. Sometimes when they are numerous enough, hunting dogs turn away the hyenas. The outcome of such encounters is never decided by strength, though. What counts is the individual nature, experience, degree of hunger, and choices that the animal has.

It seems advantageous to seize prey captured by others. But problems accompany this habit. An animal must have the ability to find and reach its prize prey quickly. It must be able to seize it and hold onto it. Scavengers with wings are best suited for the first task. They can cover large distances with only a low consumption of energy. For holding the prize, however, individual or group strength is needed. A solitary creature like the cheetah has a lightweight frame. It is especially made for running. It is not built for defending prey from others. Therefore, it is not suited for scavenging. As a result, it devotes itself entirely to hunting. The very features which make it a good hunter also make it a poor defender of its prey. So the cheetah is often robbed by scavengers.

Hunted Prey and Hunting Techniques

Of the three social carnivores, the lion, the hyena, and the hunting dog, the lion has the lowest success rate in the hunt. But it also spends the least energy at it. The lion hunts by laying an ambush and chasing its prey in short sprints. Unlike the hyena and the hunting dog which pursue their victims for many miles, it can afford some failed attacks.

The lion's success rate varies. When it hunts alone, it has an 8 percent success rate. When it hunts in groups of six or more, it has a 33 percent success rate. So why do lions ever hunt alone? The reason is simple. Twelve attempts by a lone male may result in the killing of a 90 pound (40 kg) gazelle. This may be more profitable in terms of energy expended than twelve attempts made by six lions, which result in sharing four 90 pound (40 kg) gazelles. The share would be 60 pounds (27 kg) of flesh per lion. Of course, this example applies only if certain prey is within range of one single lion. A lion hunting a buffalo alone might be gored. But a group hunt may often prove successful. In such cases, cooperation is a must.

It is hard to tell which is the lion's favorite prey. The populations of different prey vary in different places at different times. As a rule, the favorite prey is the most abundant one. In the Serengeti, the wildebeest is the animal most frequently preyed upon. Forty per cent of all kills are of wildebeests. The zebra and the buffalo are hunted less frequently. Usually the lion focuses its attention on only a few species. In various zones, three species make up 80 percent of its prey. The traditional feeding patterns of the group to which a lion belongs influences its choices.

Surprise is an essential element for ensuring success in the ambush technique. The lion approaches its prey by closely stalking it. When the prey is within reach, it darts forward and pounces. Success depends on how close the predator can get to its prey before it launches its attack. For a 50 percent chance of success, a lion must get within 36 feet (11 meters) of a wildebeest, 33 feet (10 m) of a zebra, and 20 feet (6 m) of a Thomson's gazelle.

Reliable observers believe that in some hunting situations a division of roles occurs. For instance, males do not usually join in a group hunt. They show up when the kill is over. During a group hunt, one female takes a position that will make the prey move forward. Another lies in wait and then seizes it. Sometimes the hunting arrangement is fan-shaped. It is not clear how much of this cooperative behavior is haphazard or how much is prearranged.

The kill usually occurs by suffocation and takes several minutes. Smaller prey are often killed with a final bite, cracking the backbones. The hunt usually takes place at

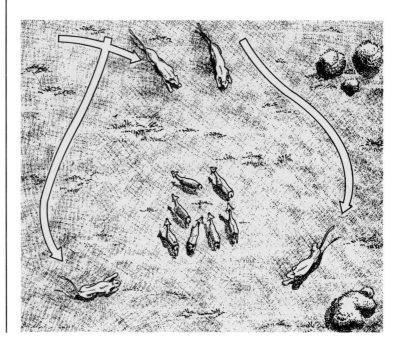

Perhaps the social interactions among lions are at their best in the coordinated hunt of the lionesses, in which one or two push the prey towards the ambush. The drawing shows how an ambush is laid between two groups. Two lionesses circle the prey, moving behind them. Two show themselves, making the group run into the trap. Lions do not take the direction of the wind into account when they plan their ambushes.

Below: The middle column of the table gives the approximate success rate percentages in hunting for each carnivore. These ratios are based on the total hunts carried out. The right-hand column compares the energy costs for hunts undertaken by the cheetah, leopard, lion, hyena, and hunting dog. Success rates and energy costs are typical of the hunting techniques used by the species. It should be remembered, for example, that the hyena is an efficient hunter. But it chooses between hunting and scavenging, depending on the nature of the situation.

cheetah	50	●●●● high
leopard	about 10	● low
lion	20	●● low
hyena	35	●●● high
hunting dog	60	●●●● high

Above: A lioness mildly threatens a cub which is lying with its legs in the air. It may have disturbed her during a meal. As a rule, lion cubs feed after the adults and do not compete with them. This feeding pattern is pitiless when food is scarce. It is the principal cause of death among the young. Regulation of lion populations is determined by this.

dusk or dawn, but nighttime and daytime hunts also take place. The lion hunts whenever the moment is right. Usually, a hunt does not last long. At most, fifteen minutes lapse from the laying of the ambush to the kill. The chase does not last more than half a minute. It will only last longer if the prey offers strong resistance. Such a case was observed once in the Ruwenzori Park. A group of lions attacked a powerful male buffalo. Before they managed to kill it, forty minutes ticked by. The buffalo is the largest animal the lion can handle. Whenever possible, lions prefer to attack isolated young individuals.

Lions do hunt buffalo. But they do not usually hunt the other members of the "big five" club, which includes the buffalo, hippo, rhino, elephant, and giraffe. They do seem to have, though, a soft spot for young giraffes. Encounters between lions and hippos are rather peculiar. On every occasion, the lions retreated. The hippos simply ignored the lions, strolling through a pride at rest, with the big cats

darting to the right and left. Observers once reported a group of lions very intrigued by a fight between hippos. They gathered around in a semicircle as if they were at a boxing match. Afterward, one of the hippos charged at them rather casually. The lions moved back a little.

Both the hyena and the hunting dog are highly social carnivores; they look different and belong to different families. They hunt in long pursuits in which they run alongside the prey, trying to wear it out or to put it at a disadvantage. Hunting dogs always hunt in groups, with most members taking part. Hyenas, however, are more flexible in their hunting techniques. Sometimes, several individuals from the same clan gather to hunt large prey. On other occasions, medium or small prey are hunted by just a few individuals. In addition, hyenas often cover long distances on their own to find a carcass. On such occasions they will certainly not ignore easy prey.

Hyenas meet about 60 percent of their feeding requirements by hunting. Their prey vary a great deal in size. A large group can kill an adult wildebeest or zebra. But they prefer young and small individuals, in spite of the opposition offered by their mothers. There are reports of female wildebeests defending their young. They have been known to kick hyenas and hunting dogs up into the air. They have pinned them to the ground with their horns, bending on their knees. But this defense only works against single predators.

In the Ngorongoro Crater, the favorite prey are, in order, the wildebeest, zebra, and Thomson's gazelle. Together they account for 90 percent of all the hyena's prey. A chart in this chapter shows the hyena's method of pursuit in hunting. Its method is quite wasteful in terms of energy used. But once the hyena starts, it has a good chance of success. Among the most common prey, it has individual success with 44 percent of adult wildebeests, and 15 percent of young wildebeests. On a group hunt, it has success with 77 percent of young wildebeests and 33 percent of Thomson's gazelles. There is a difference between the individual and group success rates when the prey is a young animal. The fierce defense by the female wildebeest clearly explains this difference.

The success rate of the hunting dog's hunt is always high, whatever the type of prey. For its size, this carnivore is capable of killing the largest prey. Its high rate of success is due to its determination as a hunter. It puts great energy into

Opposite: These drawings show moments in a lion's life. *Left, from top to bottom:* The lion is shown in walking positions, tailing, stalking, lying in wait, and pouncing. *Right, top to bottom:* The drawing shows the relaxed expression of a lioness; the alert expression of a lioness scanning for prey; the expression of a male who has picked up a scent; the tense expression of a lioness defending her dead prey; the tense expression of a male guarding his own prey. *Below:* The drawing shows the three stages of a kill by suffocation. (Redrawn from Schaller, 1977.)

Hyenas are efficient scavengers. Their powerful jaws help them break large marrow-rich bones which other predators do not eat. A highly efficient digestive system does the rest.

the hunt. In the extensively studied Serengeti, its favorite prey are the Thomson's gazelle, wildebeest, and zebra. This is true even though topi, Grant's gazelle, and kongoni are abundant. These last species are not hunted even when they are within easy reach. Why is this? It is probably a matter of learned behavior.

Hunting is more complicated than a mere chase, and the predator goes through a long and difficult apprenticeship to learn its skills. It must be very familiar with the prey's behavior, and it must be able to anticipate the prey's reactions. The development of specialized skills is advantageous. Some populations or packs of hunting dogs are specialists in hunting certain prey. In Kruger National Park, hunting dogs hunt impala almost exclusively. On occasion, even packs which are not highly specialized show sudden preferences.

A distinctive "instigation ceremony" among wolves passes on the decision to go hunting from one individual to the next. It is the leader who sets the pack in motion. Once the prey has been singled out, all members of the pack know which one is the prey. In the chase, the pack often fans out. When a prey swerves or starts to run in a curve, its path is blocked at each flank. It is as though hunting dogs are taking

A pack of hunting dogs scans the horizon in various directions in search of possible prey. These dogs have developed a very effective hunting system. In proportion to their size as carnivores, they are capable of killing the largest prey.

Following page: The leopard is currently the carnivore with the widest distribution area. It lives in Asia and Africa. This distribution does not, however, guarantee survival for the species. The leopard is hunted mercilessly outside protected areas and often within them, too. Its elusiveness and nighttime habits help it survive. Leopards often take refuge in trees. They drag their larger prey up into trees to spare them from attacks by hyenas and, above all, lions.

turns in the chase. Once the prey is blocked, it is disemboweled. The kill happens very fast. The coordination and determination of hunting dogs when on the hunt are impressive. They are known as ruthless and relentless predators. This reputation has certainly not contributed to their protection.

The leopard and the cheetah are solitary hunters of about the same size. The leopard is more sturdy and muscular. The cheetah is slimmer and taller. The maximum size of the prey that they can hunt is similar, ranging from 110 to 135 pounds (50 to 60 kg). But leopards are able to kill prey weighing up to 225 pounds (100 kg). In the Serengeti, their favorite prey is nevertheless the graceful Thomson's gazelle. Apart from this, the two animals are quite different. The leopard is active mainly at night. The cheetah is active by day. The leopard has a low success rate on the hunt; the cheetah's rate is high. A leopard's hunt will not use much energy, but the cheetah's hunt is costly. For a comparison, study the illustrated table included in this chapter on page 67.

The leopard's prey varies considerably. In the Serengeti, it preys upon twenty-four species, including such animals as jackals, birds, snakes, sheep, goats, dogs, etc. Because of its nighttime habits, little is known about many of its smaller prey. They are eaten immediately. To avoid the risk of theft by hyenas and lions, the leopard hides its larger prey in trees. Carcasses weighing over 100 pounds (45 kg) have been found in the branches several feet off the ground. There they are also safe from vultures and can be eaten over a longer period of time. The leopard hunts using the ambush technique. Ambush is followed by a very short pursuit at high speed. The leopard gets as close as possible to its prey by stalking. Then it relies on its incredible sprint.

The cheetah, which is known to all as a swift sprinter, is the fastest of all land animals. It is actually perfectly constructed for running at high speed. It has a small head and long, slender legs. When it is running fast, its backbone acts as a spring, tightening and expanding with every step it takes. Mobile shoulders enable it to lengthen its stride still farther. It runs at speeds of over 60 miles (100 km) per hour. If it came up against an obstacle or if it misplaced a foot at such a great speed, there could be dangerous consequences. Its habit of hunting only by day comes from a need for safety.

The cheetah is specialized as a sprinter. It has everything to lose by being drawn into battle with other animals. For this reason, it often does not try to defend its own prey and will give it up to intruding lions, leopards, and even hyenas. It is a rather reserved and aloof animal. It is ill at ease near park visitors and tourists.

The cheetah uses two hunting techniques. When the prey is in a large group, it walks about in full view. This puts the prey on alert. The prey stand still, uncertain what the predator intends to do. Then the cheetah breaks into a trot while the prey start to run. This is a crucial moment because the cheetah is testing the prey's energies. The selection process is not easy. Any animal that is not 100 percent fit or alert is singled out and chased.

The cheetah's second hunting technique is the surprise attack. This method works well when the cat's target is a single animal or a small group. Moving quietly through the grass, the cheetah draws as close to its prey as possible without alarming it. Then, with its great speed, the cheetah rushes upon the animal, sometimes bringing it down with one swipe of its paw.

The cheetah is very selective in its choice of prey. It shows a marked preference for young animals, which are almost always easier to catch. Here the cheetah has killed a zebra foal. This is as large a prey as it can manage. Sometimes, several individuals join together to hunt large prey like adult zebra or kongoni.

Like many predators, the cheetah prefers small and young animals. These are the most vulnerable in terms of strength, speed, and experience. The huge energy costs of the cheetah's hunt (see table on page 67) makes a high success rate important. A chase of 800 to 1,000 feet (250 to 300 m) leaves this predator exhausted. If a gazelle fools its pursuer with the right curves and side-steps, the cheetah will give up the chase. It cannot keep up a "flying pace" for more than 1,000 feet (300 m). Of twenty-six unsuccessful hunts observed in the Serengeti, twenty-three failed for this reason. Of course, the easiest prey were young animals. Of thirty-one attempts with young animals, all ended in victory.

Social Behavior of the Carnivores: The Group vs. Solitary Life

A group of lions taking their siesta in the shade of a tree shows their social behavior. The typical makeup of a pride of lions usually includes more than one adult male. Each has a well-developed mane. The pride also has up to twelve females, several young animals of both sexes not yet three

Two young cheetahs pursue a warthog, which has taken refuge in the bush. They are trying in vain to flush it out. For these predators, as for the other carnivores, the hunt is a skilled art which requires a lengthy apprenticeship. When young cheetahs leave their mothers, at about sixteen months old, they are mediocre hunters and have much to learn.

years old, and a number of cubs who are about the same age. Cubs are born to several lionesses and nursed on a shared basis.

Members of the pride do not always stay together. As the pride gathers around to eat, it is possible to watch their social structure. Adult males eat their fill first. They are followed by the females. Next, the young animals eat. Last of all, the cubs eat. When there is enough food, adult males allow cubs to eat with them. This gives the cubs a good chance of finding something choice to chew on.

A pride of lions lives in a well-defined territory. The size of this territory is based on the number of prey and the size of the pride itself. The size of animal groups in a stable environment tends to remain about the same. Nature has several ways of holding down population size. This is because a great increase in the population of a species hurts that species. In some species, only a few certain females in the group are able to give birth. This limits possible population increase. Among lions, however, all mature females give birth to cubs. This means that too many cubs are born. But their death rate is high. This is how the size of the group is limited. The young lions die without harm to the species.

This lioness and her cubs live in Amboseli Park. The females of a pride often deliver their young at the same time. So care of the cubs is shared. The survival of these latter depends on the pride's hunting ability, the abundance of prey, and the presence of adult males in the pride. If there are no adult males, nonmember males may see defenseless cubs as easy prey.

Whenever there is not enough food, they are the first to suffer. Their death stops overpopulation in a brutal but direct way.

Research sheds light on how the pride is formed. When they are three years old, males leave their own pride. They leave in small groups of two or three, almost always brothers. They live a nomadic life until they are five years of age. At that time, they battle the resident males of some other pride, fighting together as a group. If they win, they are assured of offspring.

The life of the lioness is very different. If there are losses in the group, lionesses are allowed to stay. Otherwise, at about age three, they are driven out. Then these females also become nomadic. Without a territory or a group to back them up, these lionesses have a very low reproduction rate. They may mate with an odd nomadic male, but the death

Hunting success rates for the lion in various situations (from Schaller).

Daytime hunting	21
Nighttime hunting	33
In, or almost in, the open	12
Bushes by rivers	41
A lion downwind	7
A lion upwind	18.5
A lion alone	15
Two lions together	29
Wildebeest in groups of 2 to 10	13
Solitary wildebeest	47

rate of cubs of nomadic lionesses is very high.

In the pride, adult males are interrelated, and adult females are interrelated. But the males are never related to the females. Females are expelled, and others are won. The system works to allow the size of the pride to balance with the resources available in the territory. The system also helps lions avoid interbreeding.

Among hyenas, the clan or pack is the social unit. The pack is very large, numbering forty to seventy individuals and sometimes more than a hundred. Instead of staying in one large group, hyenas form subgroups. They live in a well-defined territory. They hunt and actively defend this territory from intrusions by members of other packs. Border disputes are quite fierce. The outcome often is bloody. The hyenas' territories are quite large, but intrusions and fighting are frequent.

On one occasion, the English zoologist Jane Goodall observed a deliberate cold-blooded attack. It was led by the dominant female against two members of a rival pack, sleeping nearby. Taken by surprise, one was seriously wounded. The other barely escaped. At this point, a dozen or so of their fellow hyenas appeared. They attacked the assailants, who beat a hasty retreat. A long series of attacks and counterattacks followed. Brief raids into the territory of the rival clan were followed by rapid withdrawals.

In the animal kingdom, border disputes are frequent. But probably only the hyenas take them this seriously. Because of these bloody battles, hyenas' territorial boundaries change rapidly year in and year out. During territorial battles, some individual animals develop friendships with members of neighboring groups. This often results in genetic exchanges.

The social relations within a clan are very personal. There are identifiable social levels. But a clan gets its structure from networks of alliances, sympathies, and hostilities. The leader of the pack is an elderly dominant female. In predator societies, this is quite rare. This fact was only recently discovered because the external sex organs of males and females are surprisingly similar. In fact, until recently, zoologists thought that hyenas were hermaphroditic animals, animals with both male and female sexual features. After watching a group of hyenas, scientists discovered that they had mistaken the sex of an animal. They thought it was a large male, but it gave birth to a large litter.

The hunting dog is a carnivore which shows the high-

The drawing shows movements of a hyena on the run. Despite its apparent clumsiness, the hyena is a strong runner. It has an extremely economical run. Even its top speed is not fast, but it can still wear out its prey.

est level of sociability and cooperation. It hunts as a nomadic animal over wide areas, so it is not territorial. It lives in a group consisting of males and females of all ages and is only stationary when the young are born. Birth takes place in lairs dug for this specific purpose. The pack has an aging male as its leader. Within the pack, there are levels of dominance. However, this is seen more among the females. Only a dominant female is usually fertile.

Hunting dogs are hunters which practice their skill in very large areas. They solve the problem of bringing food back to their young by swallowing it. Later they regurgitate or bring back the stomach contents. This is a safe system which turns out to be quite handy. While the young remain with their mother, she does not take part in the hunt. So she also is fed with regurgitated flesh. This system is also used for feeding a sick member of the pack, and it works very well for young orphaned dogs. Orphans, in fact, are often adopted and fed by the group. This behavior is not typical of all animals living in groups, but the hunting dog's group acts as a particularly close unit. A few dogs take turns watching over the young animals. These dogs are fed with regurgitated meat as well. Altruism is very high. Each dog is ready to sacrifice itself for the good of the pack. The way the group acts as a unit, particularly in hunting, ensures survival for each member of the group.

When food is available, all members are equal. There are no disputes over who comes first. Sometimes a member arrives too late and finds all the food gone. If it is hungry, it is fed with regurgitated flesh. If young dogs are present at the meal, adults hold back. They let them satisfy their hunger

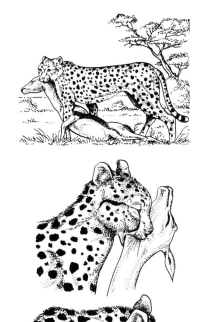

The cheetah's suffocation method of killing prey is shown.

without being disturbed. In nature, close attention to care for the young is associated with a low birth rate. It results from a clear natural law: the greater the parental care, the higher the survival rate of the young. With good care, there is less need to give birth to many young to keep group size in line with available food.

The hunting dog has perfect social organization and is an effective predator. But it is declining in number all over Africa. Among the possible causes is the extent of their hunting areas. Hunting dogs often work outside of protected zones and are subject to poaching. Their low reproductive rate is normally a well-suited adaptation. But it is not suited to a way of life that has changed drastically in recent years. Another factor is the competition from hyenas and lions. These animals are both on the rise. They both often steal prey killed by the hunting dog. Finally, hunting dogs are able to get several diseases from domestic animals. Experts agree that these reasons alone cannot explain the drop. So it is hard to decide what to do to increase the numbers of this beautiful animal.

The leopard and the cheetah are two very solitary animals. These two have a social behavior almost exclusively when they mate and rear the offspring. The leopard is territorial, but its territory is not well defined. It overlaps with others. A territory is marked with sound and smell, and individuals are careful to avoid each other. Meetings and fights between males are often bloody. But mothers, young, and siblings put up with each other. In some cases, the males help with the rearing of the young by hunting. Rare group hunts occur. They suggest that the leopard has a slight tendency toward sociability.

The cheetah is territorial like the leopard. But once in a while, it hunts migrating game. In the mating season, the courtship is catlike. The female provokes the male and then backs off. It seems that the females prefer unrelated males. This is widespread in the animal kingdom. It helps animals stay away from inbreeding. The mother gives birth to four or five young. But the death rate is high among the young. Besides disease and the possible lack of food, large predators work against the young. They attack them whenever possible. The fact that they do not have a support group to protect them means that few young survive.

Young cheetahs leave their mothers when they are about fifteen months old. By this age, they must be able to hunt on their own. The lessons taught by their mothers are

Jackal pups of the black-backed species are shown near their den. A pair of jackals is often helped in rearing their latest litter by the young born the year before. These can hunt for the pups and their mother while she suckles. They also stand guard at the den.

Two common jackal pups reply to their parents' calls by howling. Aside from maintaining contact and keeping the group together, this howling is used for delimiting a territory with "sound markers."

very useful. Initially, the young do not know how to hunt. The moment they leave their mother is a critical one. The young can survive alone only if they know their lessons well.

Jackals are delightful members of the dog family. They include the black-backed jackal, the common jackal, and the side-striped jackal. They all deserve a special mention. Many people think that jackals get their food from scavenging. It is easy to come upon them at a lion's feast. In fact, the lion, which fights off other rivals, even vultures, allows jackals to eat. They dart rapidly around the king of beasts to grab a scrap of flesh. But jackals also hunt insects and small vertebrates. They are able in hunting young gazelles. Their diet also includes some plants. Small prey are hunted alone, but jackals hunt in pairs to catch small gazelles which are defended by their mothers. Groups of black-backed jackals also join together for a short time to kill adult gazelles.

Male and female jackals form a stable couple. Bonds within the family are very strong, built by frequent greeting rituals and mutual licking. This licking happens so often

that it goes way beyond cleaning needs. A couple lives in a territory that is a few square miles in size. Relations with neighbors are polite as long as everyone stays within their own hunting territories. The young spend much of their time playing. As with other species, their games build family ties and train young for the hunt. A single family group is sometimes large. It includes the parents and young and the young born the previous year.

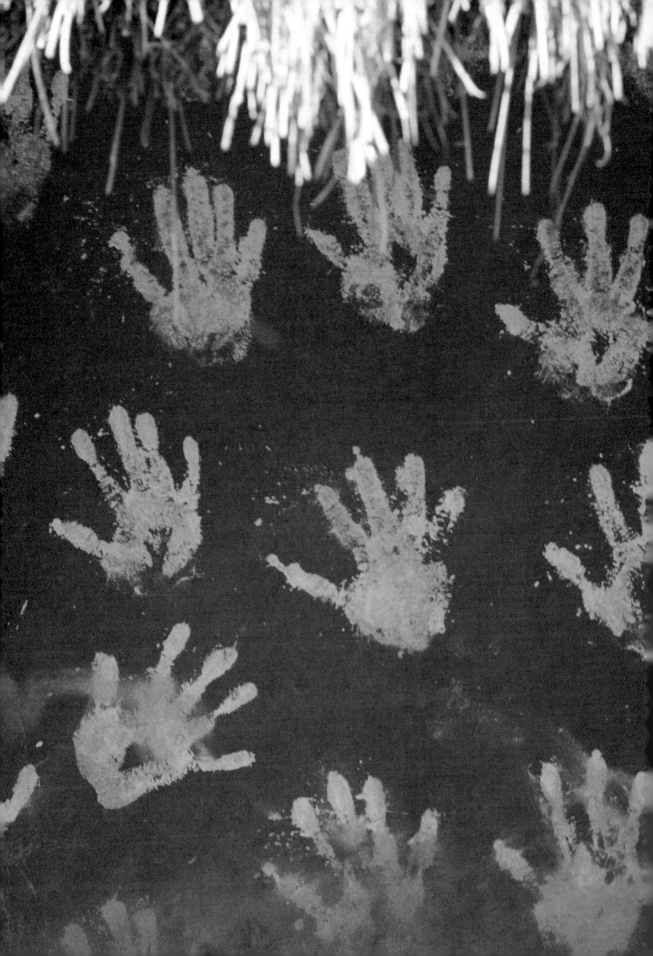

PRIMATES

Beyond the 5,600 sq. miles (14,500 sq. km) of the Serengeti and the Naabi Range lie the slopes of Ngorongoro. This is one of the most unspoiled parts of Africa. It is an expanse of grasslands. Depending on the season, these slopes are populated with herbivores which have never been confined in a park. This is the Ngorongoro Conservation Area which covers more than 2,300 sq. miles (6,000 sq. km). At times, the large migratory herds of the Serengeti roam here. The landscape has a timeless beauty. This primitive environment is wondrous in its natural beauty.

Origins

In the absence of human interference, an incredibly beautiful balance is maintained by nature. Primitive people spent millions of years in environments like the present-day savanna. There observers see an intact animal community and feel a sense of humankind's origins. For people, the savanna parks give the impression of feeling "at home." These expanses are more than just a dusty link between the Serengeti and the Ngorongoro Crater. The road map shows a detour of a few miles which leads to Olduvai Gorge. This is a deep gorge carved out by water erosion. Two million years of evolutionary history in Africa have been brought to light here.

It was here that anthropologists Louis and Mary Leakey made important discoveries concerning humankind's origins. Their fossil-hunting expeditions uncovered some of the most important fossils to date. They also found huge quantities of the oldest known stone tools. Here it is possible to study the fauna and flora associated with these finds and date the human and protohuman, or primitive, remains with some accuracy. Olduvai is the only site in the world where fossils of *Homo erectus*, *Homo habilis*, and *Australopithecus* have been found close together. So in a way, this gorge is regarded as the cradle of humankind. But the process that led to humankind's shift from apelike forms to human forms took place in a much vaster area. It took place in the age-old savanna of East Africa.

The relationship between the fossils found at Olduvai Gorge and modern mankind, or *Homo sapiens*, has not yet been clearly explained. Individuals belonging to the genus *Australopithecus* were perfectly bipedal. They were able to walk on their two feet, in an erect position, 3,600,000 years ago. They had a small brain, similar in size to that of the present-day chimpanzee. Their hands were surprisingly

Opposite: Handprints decorate the outer wall of an Acholi tribe hut along the banks of the Victoria Nile, near Gulu in northern Uganda. Humans, together with the apes, belong to a suborder of the order Primates. One of this group's characteristics is that its members have prehensile front limbs. It is likely that erect posture evolved in order to free the front limbs from their role as supports. This allowed for better handling and use of tools. A marked increase in the size of the brain and the ability to use language came after the evolution to erect posture.

similar to modern human hands. If they made tools, they used perishable materials, as no traces of them remain today. Three species belong to the genus *Australopithecus*. The species are called *A. afarensis*, *A. africanus*, and *A. robustus* (see the figure on this page). Some scientists consider *A. afarensis* as just a variation of *A. africanus*. Almost two million years ago, *Homo habilis* evolved from members of the genus *Australopithecus*. The essential feature of this new species did not involve its body form but was behavioral. *Homo habilis* had the ability to make tools. Scientists do not agree on the exact sequence or relationship between the various fossil remains. Whether it was made by the genus *Homo* or by *Australopithecus*, tool-making was a real advance in the evolution of humankind. The brain of *H. habilis* was smaller than the modern human brain. With *H. erectus*, who knew how to use fire, the size of the brain increased still farther and the tools made were greatly improved.

Exact studies make it possible to locate some human and protohuman settlements at Olduvai. Water was always a basic need. The savanna was wetter than it is now. Over the years, the savanna went through wet and dry cycles. Animal communities, richer and more varied then, supplied

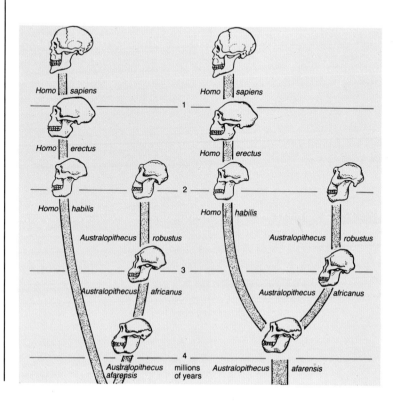

This diagram shows two of the main hypotheses on the origins of humankind. On the left is the lineage of the genus *Homo* and of the genus *Australopithecus* from a common ancestor. On the right is the direct lineage of the genus *Homo* from the genus *Australopithecus afarensis*. This species includes the famous Lucy, discovered in Ethiopia by Donald Johanson. Some scientists do not accept *Australopithecus afarensis* as an independent species. They consider it to be a variety of *Australopithecus africanus*.

A green guenon, or cercopith, perches in a tree. This species has specific alarm signals for enemies to which it is referring. An individual sends out an alarm which tells others which type of enemy it spotted. Others react to fit the situation.

Following page: A view of Olduvai Gorge is pictured. This narrow valley was carved out by the erosion of the slopes of the extinct Ngorongoro Volcano (shown in the background). It is situated in the Ngorongoro Conservation Area in Tanzania. The gorge provides a cross-section of the geological and human history of the region. It is one of the richest areas in the world in terms of human fossils. It is here that the Leakeys, after many an attempt, discovered fossil remains of *Homo habilis* and *Homo erectus*. It is here that remains of humans belonging to the genus *Australopithecus* were found, too.

plenty of food. Remains of slaughtered animals are found in old settlements. They tell how primitive humans hunted. They hunted large prey in groups, using a method based on cooperation. Two million years ago, humans had a stable home base. They hunted in groups, walked on two feet, and had the ability to make tools.

When and why did language, which is probably the basic human feature, develop? It is hard to tell with any certainty. The use of tools may have been a basic step toward developing language. Certain cooperative activities, like hunting, also may have influenced language development. Communication is essential in planning a successful hunt. Some nonhuman primates have or are able to learn the basics of gesture communications. Two primates which live in the savanna have certain special communication abilities. Communication among guenon monkeys and cooperation among baboons are worth noting.

Communication Among Guenon Monkeys

The guenon is a small monkey with a long tail. It has a gray coat with touches of olive and yellow in it and a

distinctive black face. It is often found in groups in clearings. It is always ready to escape to trees, where it is out of the reach of its enemies. It is in fact hunted by carnivores and birds of prey. It has no weapons of defense. Its life depends on how alert it is and on its ability to escape. It relies on a very good alarm system. Any member of the group who spots an enemy will warn others with loud calls. This allows all the group to find a safe place. It also defeats the enemy who loses the advantage of surprise.

Studies carried out years ago in the Amboseli Park indicate that guenon monkeys give three types of calls. These tell that a leopard, a martial eagle, or a python is near. This discovery was of great interest, but it does not mean that the guenon monkeys use certain signals to represent precise objects. It is likely that the signals sound certain levels of fright. Recently, taped recordings of the three calls were played in the wild by a loudspeaker placed close to the guenons. When the eagle signal was played, the monkeys looked upward and/or ran to take cover in the thickest shrubs. The python signal made them look around the ground. The leopard signal caused monkeys on the ground to run and hide in nearby trees.

Adults are more precise than young monkeys in their use of calls. But mistakes made by young monkeys show that they are able to tell different groups of animals apart. The young sound a leopard signal when land-dwelling animals are around. They sound an eagle signal when they see a bird. And they give a python-signal when they see snakes or long slender things. In the competition between prey and predator, this prey has made a qualitative leap forward.

Cooperation Among Baboons

When primates set out to conquer open spaces, they had to solve several problems. The first and most important was how to protect themselves from predators. When they left their tree-based homes, they became much more open to attack. This fact explains why they chose, like the herbivores, to live in groups. But among primates, the group influences the entire biology of the species. The group takes on a complex social structure.

The baboon group is formed by individuals of both sexes, all ages, and every social standing. When the group is not on the move, dominant adults keep watch. When on the move, they bring up the rear. As a rule, they are ready to run

This drawing shows primitive people trying to chase carnivores of the genus *Homotherium* away from their prey. The group hunt is one of the advantages of sociability. But another is being able to steal the kill of other hunters. This strategy is used today by hyenas and lions. It was used long ago by early humans who lived in the African savanna.

the greater risks. The social rank of the males is decided by their age. It is also decided by their physical strength, their decision-making abilities, and their ability to cooperate. Often, in fact, pairs of males defeat single males by threats alone. On a one-to-one basis, such a male might have the upper hand. Despite the occasional noisy dispute, the life of the group is fairly quiet. Serious fighting is rare. Adult males keep peace in the group. They come between young animals who become too rough in play, and they rule when low-ranking baboons start brawling.

Baboons are very caring of their young. For example, one young baboon who fell into a river was quickly saved by an adult. The young have an important social value. If a subordinate baboon has a young baboon with it, it can get close to a dominant adult. Otherwise the males chase it away. For females, holding another's young and cleaning it are very important acts. Not all mothers are willing to lend their young, however. Observers tell how one female in captivity carefully cleaned a mother before touching the young baboon in her arms. Then, almost without being noticed, she moved to the young baboon. After a while, she

gently took it from the mother. She had become so skilled at this that no mother stopped her. Other times, mothers take advantage of this attitude. They gain a baby-sitter for their young while they are busy with other things. This type of behavior reinforces social bonds. The advantages are clear when a mother dies. In this case, the young are immediately adopted.

Kinship relations between females who exchange favors are not very close. Acts which seem to be unselfish may really have selfish overtones. By handling the young for others, young females become able at handling young baboons before becoming mothers themselves. It is possible that sharing the care for the young helps build relationships between females of a group. This relationship is based on mutual benefit. As a result, a favor done today is returned tomorrow. Strong, useful ties thus can be built between adults.

emale green baboons are shown ith their young. Baboons reach exual maturity late. Females give irth for the first time when they are ur years old. They have just one ffspring. As with other mammals, e long period of suckling prevents e mother from becoming pregnant o soon again. In this way, the umber of young being born is elatively low. But protection offered y the group ensures a low mortality te.

BIRDS AND REPTILES

In the small country of Kenya there are no less than 1,033 different species of birds. This number gives a good idea of the complexity of the world of birds in the savanna. Trying to list and describe all of these birds would be an impossible task. However, it is possible to single out the most conspicuous birds and focus on how they live with other members of the ecosystem.

The different environments and changing seasons of the savanna influence not only African species. They also influence migratory birds coming from the Northern Hemisphere. These birds are nomadic, and during the dry season, they follow the winds that carry the storms. This way they find a good supply of insects, which also migrate following the storms, and new, green grass. The pattern they follow is much like the wildebeests' migratory pattern. If these birds were not nomadic, they could not exist in such large numbers in one and the same area. Their resources would be exhausted, and there would be competition from the resident species.

The Savanna Birds

The life of these birds is affected not only by the seasons but also by savanna fires. Seasons and fires give these animals special habits. Through his studies, G. B. Schaller saw birds using fire to their advantage. Marabou storks, for example, were seen pacing at the edge of such fires. Their pacing was interrupted only by occasional thrusts of their bills with which they netted mice, snakes, or other small vertebrates fleeing the flames. Lilac-breasted rollers, in their iridescent blue plumage, were also seen here, plunging into the smoke to snag grasshoppers.

A peculiar example of cooperation between birds and the other savanna animals is offered by "cleaning" birds. These birds feed on the parasites of large herbivores. The back of a giraffe may be home to several small oxpeckers. They cling firmly to its hide with their sharply curved claws. It is rare to see a rhinoceros without at least one small bird on its back. These birds search for ticks or some other blood-sucking insect. The rhino's slow movements make it an excellent perch.

Buffalo and herds of herbivores are escorted by the cattle egret, a white bird with yellow legs and beak. These birds and animals live in symbiosis, or mutual cooperation. It is not clear if the partners warn each other of predators. What might be a danger for one species may not be a danger

Opposite: A tawny eagle perches. Rather idle by nature, this bird of prey can quite often be surprised on the ground. It is often found near a carcass or perched in an acacia bush. It usually flies quite close to the ground. From the tip of the beak to the end of the tail it is 26 to 30 inches (66 to 76 cm) long. The plumage is a uniform brown color, although sometimes darker. In northern Kenya and Somalia there is a pale, cream-colored form. The tail is short and rounded. Its habitat is the open savanna, the bush and cultivated areas. It often associates with vultures around a lion's meal, but also hunts for its own prey.

Right: A hornbill is perched in a *Combretum* bush. Hornbills are medium to large-sized birds with long curved beaks. Sometimes they have a sort of horny helmet on the top of their beaks, which gives them an unmistakable look. In many species of the family *Bucerotidae*, the female is trapped in the nesting hollow during brooding. The opening of the nest is sealed up by the male with mud and dung. The bird leaves just a narrow crack open through which it can pass food to its mate.

Below: A mongoose breaks open an egg by throwing it at a stone. It grips the egg with its front paws and slips it between its hind legs. The mongoose's handling skill has been tested with a series of experiments. Tamed specimens were faced with an ostrich's egg. First they tried to break it by rolling it against a rock. When this failed, they went back to using their normal technique. But instead of throwing the egg at the rock, they threw rocks at the egg. The ostrich egg was just too big to throw.

for the other species. It is almost too idealistic to think that unrelated species act as sentries for each other.

In Kenya, studies were recently done on the relationship between the lesser mongoose, a smaller species than the common mongoose, and hornbills. Hornbills are large birds with heavy, curved beaks. They are named for the horny helmet on top of their beak, which doubles its size. Hornbills follow mongooses when they move as a group in search of food. While exploring the ground, mongooses unearth winged and jumping insects. These make a handy mouthful for hornbills. Mongooses, on the other hand, use hornbills as their sentries. Hornbills sound the alarm whenever a bird of prey draws near. Hornbills also give warning when they spot species which prey only on mongooses. This suggests that their cooperation has been going on for a long time. Sometimes the mongooses even delay setting off in search of food if they do not see their company of hornbills. The hornbills, in turn, wait for the mongooses. They perch in trees around termites' nests where mongooses live. If mongooses show no signs of life, the hornbills wake them up. They fly close by and poke their heads into the ventilation chimneys of the termites' nests.

A group of ostriches lives in the Ngorongoro Crater. Behind them, some wildebeest move away. Male ostriches have black-and-white plumage. Females and young have a less colorful plumage, somewhere between brown and gray. From birth to the age of two months, the chicks are white or reddish. They have black markings on the head and neck. When sitting on the eggs, the females look like mounds of earth. The male is much more visible. He does not take part in the work except during times when his plumage is less noticeable.

Landlubbers: Ostriches and Bustards

In arid grassland areas where the Grant's gazelle and the oryx live, a group of ostriches appears on the horizon. It is impossible to get close to them because they run off at the least disturbance. In May, as the mating season draws near, the male's neck and thighs become red. During the courtship ceremony, males lay flat on the ground with their wings wide open. They swing their bright red heads and necks while they make dull, bellowlike noises. Meanwhile, females circle around them. They flutter their wings and lower their heads as if performing a dance. The male establishes a territory of about 6 sq. miles (15 sq. km). There he digs a hole in the ground which will serve as the nest. The senior female of his harem is the first to mate and the first to lay eggs. After a few days, other, minor females are added to the mating list. Minor females are females who have failed to find a mate or have somehow lost their nest. They share a male and share a nest in order to insure that they have offspring. Minor females use the nest of the dominant female to lay their eggs, but they do not brood them. This is left to the senior female and to the male. The male broods

only early in the morning or late in the afternoon, when bright feathers are less visible.

Sometimes minor females are not welcome. It has been observed that a dominant female rolled some of the eggs laid by other females out of the nest. The nest contained thirty to forty eggs, too many even for an ostrich. The dominant female carefully kept her own eggs in the middle. The eggs at the edge are the most vulnerable to predation by jackals and vultures. Thus, the dominant female takes advantage of the presence of others. She shares guard duties, and she thins down the other eggs in the common nest. This way her eggs are more likely to hatch. In fact, all of the dominant female's eggs hatch. But only about half of the eggs laid by the other birds hatch.

Another colorful bird of the grasslands is the great bustard. The bustard is smaller than the 6 to 7 feet (2 m) tall ostrich. It grows to a height of about 3 feet (1 m) and can weigh more than 22 pounds (10 kg). The male is territorial. He has the unusual ability to change his appearance during the courtship ritual, held in the short rainy season. He does not normally have colorful plumage; his black crest, long

A dominant ostrich works on her nest. The nest is a hole in the ground usually dug in the lee of a bush or in the grass. The dominant female and the male are the only ones who sit on the eggs. She rolls eggs laid by lesser females in the harem to the edge of the nest. There they are more likely to be eaten by enemies. Or they may spoil from too much sunlight. An ostrich egg is tasty prey, especially for lions, hyenas, and vultures. It equals about twenty hen eggs. The nesting systems of the ostrich save at least part of each clutch.

A weaverbird perches on an acacia branch. The forest species are usually insect-eaters. They search for food individually. They are monogamous. One male mates with one female. They are territorial and build hidden isolated nests. The savanna species tend to lead a sociable existence. They feed in flocks on seeds and build their nests in the branches of a single acacia, in a colony. They nest in the savanna but also in cultivated areas, close to human settlements. Some species of weaverbirds gather in flocks sometimes numbering more than a million birds, which explains clearly how they can lay waste a field of wheat in a matter of hours.

gray neck, and dark-colored back are not very showy. But when he is strutting about his territory, his gait becomes slow and measured. His tail is held erect, and the white coverts, which are the feathers beneath the tail, shine in the sun from a long way off. He puffs up his neck and white sac at the throat. By opening and shutting his beak in quick succession, he makes a noise that sounds like a drum beat.

Nesting Problems: The Weaverbirds

Among acacia branches, there seems to be an unusual abundance of large, round fruit. Up close, this fruit turns out to be carefully woven nests. The weaverbirds of the Ploceidae family build these domelike nests. The nests have a door at the side, at the top, or at the bottom.

The weaverbirds belonging to the genus *Ploceus* are yellow and black in every combination. Some have a dark mask against a light-colored body. Others are the very opposite. Members of the genera *Malimbus* and *Euplectes* are colored with a touch of scarlet or orange. Among these birds, the variety in social organization is astonishing. Some species are solitary, others sociable. Some have couple

Three vulturine guinea-fowl perch on a tree limb. These colorful birds have a necklace of blue-black lancelike feathers with a lengthwise stripe. The rest of the plumage is black with white speckling. The head and the neck are featherless and gray, with a patch of brown feathers on the nape. The head is small compared with the rest of the body. The elongated neck gives this guinea-fowl the look of a vulture. It lives in flocks and in arid bush. It is widespread in Somalia, eastern Kenya, and Tanzania. The arrival of the rains in November marks the start of the mating season for many bird species, in particular those which nest on the ground. If the short rains are late, however, the guinea-fowl remains in its flock. It puts off the choice of a mate and reproduction until April when the grass is tall again.

The Egyptian vulture is an unusual looking bird. It has bare yellow skin on the head and throat and a white, hairy collar. It opens ostrich eggs by hurling stones at them. It picks up the stones with its beak. It can hurl them either while perched or in flight. Other vultures may be larger in size. But they cannot break the thick ostrich shell using just their beak and talons. So they in turn prey on eggs opened up by the Egyptian vulture. This vulture also cracks smaller eggs open by dropping them on the ground.

relationships, others have a harem pattern. Some build concealed and isolated nests, others build nests all together in brightly-colored colonies. Evolution experts say these differences result from changeable factors in the environment. Factors like type of food, its distribution and abundance, and predators affect the birds' social structure.

Species of weaverbirds which live in thickly wooded areas, in dense bush, or in forests along rivers, tend to eat insects. The search for food there makes the pair a strong unit. The couple must defend a large territory in order to ensure a sufficient amount. In this territory, the birds will also build a hidden nest. Catching insects requires a lot of energy. So both the male and the female tend to the young, frequently visiting the nest. Their coloring is quite dull and is similar in both sexes. This is a camouflage device, called "mimetism," used to reduce vulnerability.

Other species which live in the open savanna eat seeds. They feed in flocks and form huge nesting colonies. In the grasslands, it is hard to hide a nest. So they mass together in the rare spiny acacia trees. There the density of their nests helps to provide a little shade from the sun. Within a colony, males compete to offer the most attractive nesting position to females. Because food is plentiful, males do not care for the young. They have a tendency toward polygamy, so one male may mate with many females. Males have a splendid plumage, which they use to attract the females.

Birds of Prey: Scavengers and Hunters

Among the carnivorous birds, it is mainly the vultures that feed regularly on carrion. Among the eagles, only the tawny eagle eats like the vultures. It eats carrion as an alternative food at certain times of the year. It is not easy to group the large carnivores as hunters or as scavengers, since they are often both. The task is much easier with birds. The term *scavenger* in no way implies anything negative. Eating with the lion is sometimes very risky.

Scavengers are grouped in categories according to their feeding strategies. Small Egyptian vultures will only pick up the crumbs from other animals' meals. They usually prefer eating termites, reptiles, and eggs. The Ruppell's vulture and the white-backed vulture, both griffon vultures, are large and heavy. The white-backed vulture weighs 11 pounds (5 kg). The Ruppell's vulture weighs more than 15 pounds (7 kg). It has a long, sharp beak, and a distinctive, slender, and almost featherless neck. Its diet is based

A group of griffon vultures gathers around a carcass, probably of a small wildebeest. In the background stands a marabou stork, waiting for its turn. Both the griffon vulture and Ruppell's griffon are common throughout East Africa, and in large reserves in particular. The griffon vulture nests in clumps of trees along streams or in forests. Ruppell's griffon nests in rock walls. Both species live in large groups. They are not territorial but, instead, follow the movements of large migratory herbivores.

entirely on carrion. Such a fussy diet can evolve only if large numbers of animals with a high death rate are available. This is the case for the migratory herds of the Serengeti. No matter where or when an animal dies, its carcass is spotted and fed upon by large numbers of vultures. Griffon vultures are not territorial. They follow migrating herds of herbivores, covering great distances and flying high in the sky. They are always ready to swoop down at speeds of up to 45 miles (70 km) per hour when their acute eyesight spots a dead body on the ground.

Left, top: The drawing shows a griffon or white-backed vulture. A white band on the front edge of the wings gives the griffon its typical shape in flight. *Below:* The drawing shows a Ruppell's griffon. This griffon has white markings on the back, breast, and wings. Three narrow pale bands of white on the lower surface of the wings are visible in flight. *Top, right:* The white-headed vulture has a blue-and-red beak. It has a white head and abdomen and three white bands beneath the wings. *Below:* The eared vulture is the largest of the African vultures. It has folds of bare skin on the head and a white stripe beneath the wings. Vultures have very large wingspans, short tails, and bare heads, except for the lammergeier. Males and females look alike. They nest in rocks or trees.

The griffon vultures' habit of searching far and wide for food is very well suited to the distribution of the resources in the Serengeti. The griffon vulture is the most important flesh-eater in the ecosystem. Together, all carnivorous mammals eat fourteen thousand tons of flesh a year, while the griffon vultures alone consume eleven tons. In areas where herbivores are residents, as in the Ngorongoro Crater, hyenas and lions are the primary predators. There is little left over for vultures. So few vultures live in this area. In the Serengeti plains, however, land-dwelling carnivores take advantage of the airborne scavengers. By following their flight, they can reach a carcass just minutes after the vultures. They quickly chase them away.

The white-headed vulture is one of the few which also hunts. It has a different strategy. It is territorial and prefers to single out small animals such as rabbits. It knows its territory by heart, and it does not like to attend great banquets around a carcass. It likes to eat without too many fellow diners, or rivals, around and will fly off at the slightest threat.

The eared vulture is a territorial species. It is very aggressive and will chase griffon vultures away from a carcass, mainly because of its intimidating size. It does not reach the 15 pound (7 kg) size of Ruppell's vulture. But its wingspan is almost 10 feet (3 m), and it seems larger during its displays of force. Rarely do more than six gather around the same prey. With their large, bulky beaks, they rip away shreds of hide, flesh, and tendons. They eat like the lammergeier or bearded vulture does. The lammergeier vulture is the only species with its head and neck entirely feathered. The lammergeier arrives late for meals. It eats the bones, flinging them at rocks and then gathering up pieces and marrow. It is also territorial. Each pair controls a large area, surveying it with the soaring flight typical of all the vultures. This type of flight makes use of the rising currents, so it is very energy-saving. It requires about one-thirtieth of the energy needed in flight using wing beats. This adaptation enables vultures to easily follow columns of migrating animals or to patrol an area for a whole day.

It is possible to tell vultures which perch haphazardly in trees from those which focus on prey, waiting for the lions to leave it alone. By following the vultures' flight, it is easy to reach the site of a hunt. Around a carcass there may also be marabou storks with their bright pink throat sacs. With an

A bearded vulture or lammergeier is the only vulture which has a head and neck that are completely feathered. The head is cream-colored. It has a black marking running from the eye to the black tuft beneath the beak. The breast and legs are covered with an orange-colored plumage. They contrast with the dark wings. The shape in flight is more like that of a huge falcon. It has long, narrow wings and a long, dark, wedge-shaped tail. Normally alone, it nests in caves of crags.

Above: The bateleur eagle, a bird of prey, often adopts the strategy of a scavenger. It is fairly common in the savanna and scrub of central-eastern Africa. *From bottom to top:* A female, with wings spread, and a male, perched; the male and female in flight.

Opposite: A secretary bird battles with a snake. This species is found only in Africa and has a distinctive crest of black feathers on the back of the head.

awkward four-legged gait, they circle in search of morsels of flesh. Their long, storklike beaks are not designed for stripping flesh from bones, so they steal a few mouthfuls from vultures and others. At times they do not manage to catch any fish or frogs. Then they live on the garbage close to villages. They resemble vultures when they fly. As they sail through the air, they hold their heads back.

Birds of prey are not an important part of this ecosystem. But they create images which are hard to forget. The tawny eagle may join vultures around carrion. It has a generally idle life-style. One may be found on the ground, near a prey. More rarely one skims close to the ground.

The martial eagle is bigger. The female weighs up to 11 pounds (5 kg) as opposed to the tawny eagle's 6 or 7 pounds (3 kg). It has brown plumage, with pale legs and belly. As a rule, it perches at the top of an acacia tree. It can lift baboons, gazelles, and even small impala off the ground. But it does not always get the better of vultures. A martial eagle which grabs a black-backed jackal can be attacked in flight by an eared vulture and forced to land and abandon its prey.

Certain smaller kites, like the African black kite, are no less daring. In villages, they steal bones from dogs. They make nosedives to scatter a group of feeding jackals. They prey on small mammals or birds. They also feed on carrion, gathering in flocks to do so. The black-winged kite appears in grasslands after plenty of rain. It eats mice of the genus *Arvicanthus* which increase in the thick green grass. This is a hunting kite whose shape looks very impressive silhouetted against the sky. It hovers into the wind on its long pointed wings. It suddenly plummets down to the ground for a kill. But it is the bateleur eagle which provides a real show. Its bold acrobatics are performed during the mating season. Its wings are covered with long feathers. During display, they produce sounds which can be heard a long way off. If this eagle flies off with a hat, it can make it spin in the air, then catch it again like a juggler.

A Word About Reptiles

The python is much smaller than reported in the journals of travelers. The largest creature the python can attack is a small antelope. Here the myth surrounding the python overlaps with the myth of unexplored Africa to produce small masterpieces of zoological fantasy.

Myth and age-old fear of the snake are shared through-

The common night adder is a ground-dwelling adder of the central African savanna. It usually lives close to water. From December until the first rains in March, it hides and is very hard to find. The females lay twelve to twenty-six eggs in December. These hatch after fifteen weeks, during the rainy season. This snake has well-developed poison glands. Its bite is not very dangerous, at least not for humans. It has a rhombus-shaped black marking on the head, between the eyes.

out the world. Many people fear snakes. This reaction is also an innate feature of other primates. When monkeys and apes sight a snake, they become alarmed and panic. Perhaps monkeys are innately afraid of snakes simply because snakes are dangerous.

In the savanna, two adders, the puff adder and the Gabonese adder, exist. *The Guinness Book of World Records* places these two at the top of the list for having the longest fangs of any snake. Apparently, they can kill an elephant. But this may be just another snake myth. Certainly, a large male leopard found dead with signs of a snakebite on a front leg is evidence of the danger of these snakes. Adders are the size of a motorcycle tire and about 6 feet (2 m) long. They are slow and sluggish but very speedy when they strike. They are active at night, hunting rodents and small vertebrates. The puff adder is so called because, if disturbed, it makes loud warning hisses which sound like puffs. Adders, like all snakes in fact, attack only large vertebrates for defensive reasons.

Small carnivores like the jackal, long-eared fox, or serval can easily kill their attacker, if bitten. So it is vital that adders carefully choose the targets of their attacks. Snakes

The agama is probably the easiest reptile to spot in the savanna. The male controls a territory and up to half a dozen females. During courtship he approaches the female swinging his head to and fro. To defend his territory from rival males, he also makes signals with his head. Two males nod repeatedly with the same jerky movements. Only mature males strong enough to outdo others keep their bright blue-red-orange coloring. Subordinate males have a dull brown coloring.

are part of the jackal's diet, but the jackal can tell poisonous species and carefully avoids them.

Two other important poisonous snakes can live in the savanna. They are the black-necked cobra and the black cobra. The black-necked cobra spits its poison up to 6 feet (2 m) away. It aims at the eyes and blinds its prey or attacker. The black cobra is more traditional and lands a fatal bite. These snakes live among dense vegetation and in tall grass. It is possible to find them at night. Otherwise, the easiest way is to buy a ticket for the Nairobi Snake Park and visit the reptile collection.

The friendly agama is easy to find. It has an apparent liking for the stones around bush lodges. This large lizard can reach a length of 12 inches (30 cm). The male is territorial and polygamous. It has unmistakable coloring with its red head and bright blue body. Its colors and displays make it easily visible to other males and females. This is, in fact, how it signals its presence and marks its territory.

GUIDE TO AREAS OF NATURAL INTEREST

Few game reserves and national parks established to protect the savanna environment come close to being self-sufficient ecological units. In spite of their size, they are usually too small to contain natural migration, grazing, and predation cycles. This raises the problem of how to manage these areas if, for example, a particular species grows too numerous.

Tourism is an external factor in the ecosystem that must somehow be controlled. Too many roads and too many minibuses drive through the grass. Too many safari lodges and too many staff villages alter the appearance of a park and affect animal habits. The roads become migration routes for herbivores. Swimming pools at lodges become alternative water holes in the dry season. Garbage becomes an inexhaustible storehouse for hyenas, jackals, and marabou storks. These gather in large numbers around some lodges with unforeseeable consequences. However, when controlled and well managed, tourism is not in contrast with the management of parks. A trip in the savanna can be a fine experience for anyone looking for a natural environment which has retained the look of far-gone ages.

Protected areas listed in this chapter represent typical, easy-to-visit examples of savanna. They offer various accommodations such as safari lodges, which are comfortable hotels often organized in the form of cottages. Campsites are simply areas where camping is permitted; they are not equipped with facilities. Visitors wanting to hire their own vehicle should know that four-wheel drive is needed only in certain places. It is needed to drive down into the Ngorongoro Crater or for visits during the rainy season. Also, hiring a vehicle in Kenya is much cheaper than in Tanzania.

The principal factor which makes a trip to the savanna a good one is the weather. Coming upon a hunt by following a flight of vultures might take several hours. But it is a vastly different experience than hurrying along with many other minibuses to where a pride of lions is having lunch. Animals must be approached gradually. Visitors must not get too close to them. There is a fixed distance at which they will take flight and pursuit will be of no use. In addition, flight stresses animals. The best way to observe savanna life is to wait until animals will become unaware of visitors. The time required for seeking out animals and for patient waiting gives the day a different rhythm. Activities are based mainly on the position of the sun.

Opposite page: A hot-air balloon safari rises over the Masai Mara Park in Kenya. The Masai Mara is a relatively small reserve compared to other parks. It is 700 sq. miles (1,812 sq. km.) in size. But it is well organized. A third of it is under full protection, with no human settlements allowed. Possible trips within the park include a balloon safari which lasts several hours. It starts at dawn from the Keeborok Lodge, in the heart of the most protected area. The balloon trip provides extraordinary views. But the balloon also offers chances to get close to animals without making noise and alarming them.

Right: The savanna parks and areas where the savanna is well represented range from west Africa (the "Guinea savanna") to central-eastern Africa (the vast grasslands and the mixed areas of "Miombo" forest) to the savanna and bush of South Africa.

Below: This area is covered by savanna environments in the continent of Africa.

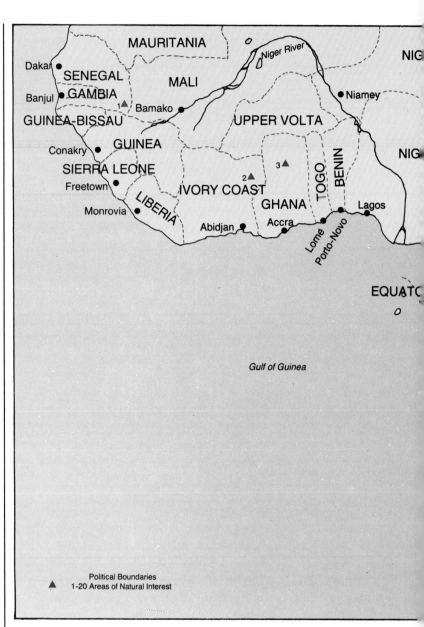

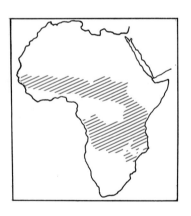

SENEGAL, IVORY COAST, GHANA, NIGERIA, CAMEROON, SUDAN

The so-called Guinea savanna is a belt of thickly wooded grasslands which stretches from Senegal to Sudan. To the north it gradually turns into the Sahel, which is dry land with very scarce vegetation and desert. To the south it becomes rain forest and connects with the savanna of East Africa. This environment has grazing grounds of tall grasses of the type *Panicum maximum* or "Guinea grass." It plays host to numerous nonmigratory herbivores. Among them

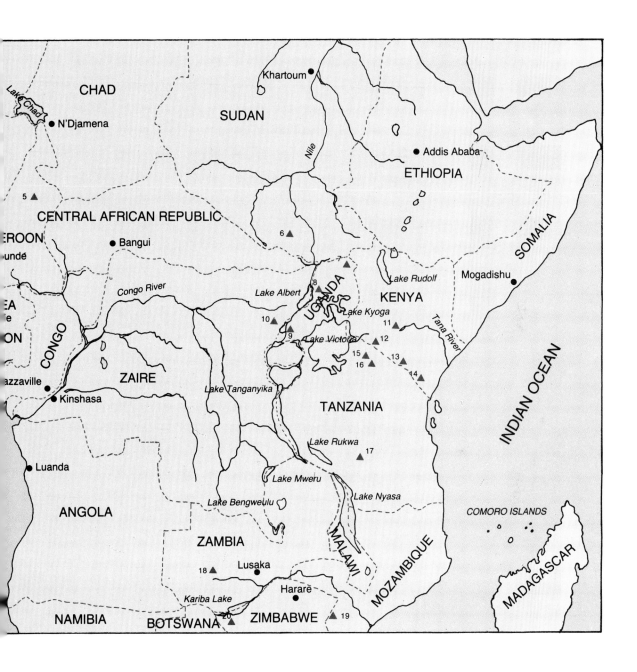

are the massive greater eland, named after Lord Derby, the bushbuck, and the roan antelope. All the large carnivores also live here. Many animals show a certain melanism, or a darkening of the skin and hides, which is common with forest living. There are no zebras or black rhinoceroses, except for a small population in Cameroon. But on the edge of the Sahel there are white rhinos. Below is a list of parks where it is easy to find good facilities.

Niokolo-Koba (1)
La Comoe (2)
Mole (3)
Yankari (4)
La Benoue (5)
Southern N.P. (6)

Opposite: The Kabalenga Falls are situated in the heart of a park of the same name, in Uganda. The Victoria Nile at this point flows into a crack in the rock hardly 20 feet (6 m) across. From there it drops more than 130 feet (40 m). It is possible to make a boat trip to the foot of the falls. This trip crosses an area rich in birds, elephants, buffalo, waterbuck, and the largest concentration of crocodiles in Africa.

UGANDA

Kidepo Valley (7)

Kabalenga Falls (8)

Senegal: Niokolo-Koba National Park is about 3,100 sq. miles (8,000 sq. km) in size. Plenty of buffalo, roan antelope, Uganda kob (waterbuck), and greater eland roam here.

Ivory Coast: La Comoe National Park is about 4,500 sq. miles (11,500 sq. km) in size. It features large populations of buffalo, hippopotamuses, and roan antelope.

Ghana: Mole National Park is about 1,500 sq. miles (3,900 sq. km.) in size. Numerous kongoni and Pata monkeys live here.

Nigeria: Yankari Reserve is about 780 sq. miles (2,000 sq. km.) in size. It features populations of roan antelope and kob.

Cameroon: La Bénoué National Park is about 700 sq. miles (1,800 sq. km.) in size. It has a large population of red, or Pata, monkeys.

Sudan: Southern National Park is about 6,500 sq. miles (16,800 sq. km.) Greater eland, kongoni, roan antelope, and white rhinoceros live here. There are no tourist lodges.

The Kidepo Valley Park sits on the northeast border of the country. The 490 sq. mile (1,260 sq. km) Kidepo Valley Park is dominated by Mount Morungole and the wooded ridges of Mount Lotuke, in Sudan. These two peaks rise to almost 9,200 feet (2,800 m). The park is crossed by the Kidepo and Laurus rivers, which are dry for most of the year. It is typical arid savanna land with large areas of acacia and palms of the genus *Borassus*. The best period to visit it is between December and early April, in the dry season. During the rest of the year it can be difficult to get about, even with a four-wheel drive vehicle. This environment is half way between savanna and the Sahel. Bright's gazelle, a subspecies of Grant's gazelle, lives here, and is well suited for surviving periods of drought. Other species which are able to live in semidesert zones also live here. They include the Rothschild's giraffe, a species with very dark markings in star-shaped patterns and white legs below the knee. There is also Kirk's dik-dik, which is one of the smallest dik-dik species, the eland, the lesser kudu, and the cheetah.

The 1,544 sq. mile (4,000 sq.km) Kabalenga Falls Park includes the Kabalenga Falls, once called the "Murchison Falls," which gather the waters of the Victoria Nile. Very heavy rains in 1961 opened up two smaller falls alongside the main falls. They make this sight even more spectacular.

Most of the area is covered by rolling grasslands. The grasslands are interrupted by strips of river forest, stretches of wooded savanna, and a few isolated forests. These are home to chimpanzees.

The grasslands are alive with elephants, which have increased in number beyond all expectations. They now number more than eight thousand head. They are being slaughtered to control the population growth. In addition, this area has the largest concentration of crocodiles in

A group of elephants, adults and young, grazes in a wooded part of the savanna. The foreground shows tree destruction caused by elephants passing through the landscape. The tree here has been stripped of much of its bark. When eating, elephants do not select any species or size of tree. Trees under 3 feet (1 m) tall only rarely become part of their diet. It is quite common to surprise a group busily feeding on grass, leaves, or bark. Elephants do not chew the cud. They are not ruminants, so they do not get much nourishment from the food they swallow. They spend about eighteen hours a day feeding. They make up with quantity for the limited efficiency of their digestion process. Elephants are very voracious. They consume an average 310 to 400 pounds (140 to 180 kg) of forage a day.

Africa, plenty of hippopotamuses, buffalo, kob, and Rothschild's giraffe. The white rhinoceros also has been reintroduced here. Outcrops of salt exist here and there. Many animals are very fond of this salt. Once these outcrops were used by safari hunters. Today they make good vantage points for observing the wildlife.

Ruwenzori Park used to be called "Queen Elizabeth Park." It covers almost 780 sq. miles (2,000 sq. km) inside the

Ruwenzori (9)

Rift Valley. Its altitude ranges from 3,300 to 5,000 feet (1,000 to 1,500 m). To the northwest it is dominated by the snowy peaks of the Ruwenzori Mountains. These mountains are identified with Ptolemy's Mountains of the Moon. Their peaks soar to heights of nearly 16,400 feet (5,000 m), and extend toward the park's boundaries. This area is covered with ancient volcanoes and is known as the "explosion zone." Here the landscape is very peculiar. It features crags and green or blue lakes hidden in the craters. In the southern part, savanna plains are populated with large herbivores. These include the topi, found only here in Uganda, the Ugandan kob (or waterbuck) and buffalo. The reddish coloring of the buffalo hides make them look like a result of crossbreeding with the forest-dwelling buffalo of Zaire. The Kigezi district is famous for the lions' habit of climbing into trees. They do the same at Lake Manyara National Park, in Tanzania. Despite the general abundance of animals, there are some unexpected absences. There are no rhinos at all, and the same goes for giraffe, zebra, and kongoni. The Maramagambo Forest is a sanctuary for the chimpanzee, the baboon, and the colobus monkey. The monkeys are easy to hear but not so easy to spot.

The animals in this forest are renowned for not being afraid of people. Hippos are usually timid and move about only by night. But here they are active in broad daylight between the lakes and the mud-holes.

ZAIRE

Virunga (10)

Virunga Park covers 3,100 sq. miles (8,000 sq. km) and is an historical landmark. Founded in 1929 as the Albert Park, it was one of the first national parks established in Africa. To the southwest stand some active volcanoes and scattered trees. This is home for chimpanzees, hyraxes, and forest buffalo. The buffalo have longer, redder coats than those living in grasslands. They are also smaller and have more curved horns. To the southeast the volcanoes are extinct, and a dense green bamboo forest grows in the lava soil. This forest hides the mountain gorilla. Wide plains open up in the middle of the park. Here is Lake Edward, densely populated with a large variety of birds and hippopotamuses. But it has no crocodiles. To the north, the savanna becomes more wooded. There are fewer antelope, but large numbers of buffalo, elephant, and lion live here.

Semliki River links lakes Edward and Albert. Along the river, now called "Lake Mobutu Sese Seko," there are virtu-

KENYA

Meru (11)

Masai Mara (12)

Amboseli Masai (13)

ally impenetrable areas of tropical jungle. The carnivore population includes the lion, the hunting dog, the leopard, and the aardwolf. There are good facilities around Rwindi. It is advisable to have four-wheel drive for getting about.

The 700-square-mile (1,800 sq.km) Meru Park lies to the northeast of Mount Kenya. It is crossed by several streams, making it green and wooded. To the north, there are shrubs of the genus *Combretum*, palm groves, and a corner of rain forest. This area is well known for the book *Born Free* by Joy Adamson about the lioness called Elsa. There is a camp named after her on the Ura River. This is a roadless and completely wild area where it is possible to hike and camp with a ranger as escort. The area is home for the white rhinoceros, which is only found here in Kenya.

The Masai Mara area covers some 700 sq. miles (1,800 sq. km) and is part national park and part game reserve. In other words, the land may not be used in any way. Human settlements are forbidden. A distinctive feature of the landscape is the green hill country and the river forest along the Mara, the Talek, and the Sand rivers. The largest lion population in Kenya and a large concentration of topi and antelope live here. These include the roan antelope, with its long, black, scimitar-shaped horns. It is not found in any other Kenyan park. Excellent facilities are available at the Keekorok Lodge, in the heart of the park, and at the Mara Serena Lodge, built on a hill. Along the Mara River there are two well appointed camps, Wildlife Safari Camp and Governor's Camp. But there is also splendid, inexpensive camping on the Sand River, on the border with Tanzania. Not far from the northern entrance, on the road to Narok, is Cottar's Camp. This is a flower-filled oasis where visitors are able to set off on dawn and dusk safaris on foot. This is one of the rare opportunities to walk through the savanna with expert and armed guides. Visitors can also spend the night in the look-out, a cabin for observation built in the middle of the savanna.

Amboseli Masai is a nature reserve of some 500 sq. miles (1,250 sq. km). The term reserve means that there is overall protection of both flora and fauna. But the Masai, a pastoral tribe wearing distinctive purplish red tuniclike robes, live in the area. They graze their livestock in it. This is

Buffalo are shown at a watering hole in the Tsavo Park in Kenya. The stripes of green grass indicate the presence of surface water and follow its winding course. The rare trees and green bushes in the generally gray panorama belong to species which do not shed their leaves. They stand out sharply in the dry season. The eastern part of the park is the most arid and flat. An average 16 inches (40 cm) of rain falls there each year. In the western region, around the lush Chyulu Hills, the average yearly rainfall is 78 inches (200 cm). With the first light rains, savanna bushes flower briefly. They totally transform plant life which seemed to be dead.

an open lowland area fringing Lake Amboseli. It is scattered with acacia groves. It is dominated by the view of Mount Kilimanjaro, which is always snow-capped.

In the eastern part of the reserve, it is possible to see many animals in just one morning. The elephant, black rhino, lion, cheetah, or the Masai giraffe, a subspecies with two sets of horns and markings with irregular edges, live here. The western part is more arid. Both gerenuk and beisa live here. With a bit of luck, it is possible to surprise the long-eared fox, also called the "bat-eared fox," spread-eagled in the sun near its den.

The bird population is rich and varied around the lakes. It includes among the most eye-catching species, the *Falco fasciinucha* and the *Ploceus castaneiceps*. The first has a bright plumage and looks like a lanner falcon. The second is a species of yellow weaverbird with orange splashes. It is only common within the confines of this reserve.

Tsavo (14)

Tsavo National Park was established in 1948. It includes a large number of environments in its 7,700 sq. miles (20,000 sq. km). Parts of the park are inaccessible because of a lack of roads. There are open grasslands, savanna with bush, palm groves, mountain forest, semiarid zones, and desertlike steppes. Around the Chyulu Range, to the west, recent volcanic activity has marked the landscape with lava flows and cones. Shitani Peak stands near the Kilaguni Lodge. In this volcanic zone are the famous Mzima Falls, home to numerous hippopotamuses. The water here is so clear that it is possible to follow the hippos' movements through the portholes of underwater observation chambers.

Another major show is offered by the watering holes. There in the dry season, hundreds of elephants come to drink and bathe. Mudanda Rock, an observation post just above water level, gives a really close-up view of their activities. In the bush along the Galana River lives the lesser kudu. This is one of the most beautiful antelope, with spiral horns and a striped coat.

TANZANIA

Serengeti (15)

Serengeti National Park was established in 1952. It covers an area of 5,600 sq. miles (14,500 sq. km). The Masai Mara and Ngorongoro reserves are adjoining the park, forming a large unit. It embraces many environments, from river forest to grasslands with kopjes. It hosts an impressive concentration of herbivores and carnivores. In just a single day it is quite common to see forty or more lions. Even the leopard is relatively common along the Seronera River. In addition, all the species of jackal, the common, black-backed, and side-striped, live here. The area also is home for the aardwolf. The bird life is just as rich. Without looking too hard, it is possible to see flights of colorful rollers, kingfishers, and sunbirds with their long slender curved beaks. Close to the Seronera Research Center there is an important site for savanna ecology studies. It is the Seronera Lodge. It is built on a kopje whose huge granite boulders are part of the walls and furnishings of the hall. At Lake Legaja where Jane Goodall worked, there is a permanent campsite called "Ndutu Tended Camp." It is well worth staying at, not only for its beauty, but also for the pictures left there by the various researchers.

Ngorongoro (16)

Ngorongoro Conservation Area covers 2,500 sq. miles (6,500 sq. km). It includes the volcanic Ngorongoro Highlands and Olduvai Gorge. It is almost obligatory to stop at Olduvai Gorge to see the famous fossil finds of primitive humans. A small but well-organized museum there illustrates this history. Moving from the dry Olduvai area, high altitude forests cover the slopes of Ngorongoro. Green grasslands cover the bottom of the large crater which, because of its size, is called a "caldera."

Here there is a high concentration of animals all year round. The big five—the elephant, rhino, hippo, giraffe, and buffalo live here. Lion, cheetah, leopard, large numbers of side-striped hyenas, and the African hunting dog also live here. Among the bird life, the lammergeier vulture and Verreaux' eagle both nest on the crater's slopes.

Ruaha (17)

Established only in 1964, Ruaha National Park covers about 5,000 sq. miles (13,000 sq. km). It lies along the course of the Njombe and Great Ruaha rivers. During the dry season between June and November, large groups of elephants, buffalo, impala, and giraffe live near the rivers. Here it is also possible to come upon the greater kudu. It has spiral-shaped horns and a massive, striped body. This is perhaps the only part of East Africa where tourists can almost certainly take its picture.

The landscape is given its special character by the "Miombo" forest. This is a dense grove of deciduous trees growing in a tall grassland. Here live the roan antelope and

Following page: The sun sets over baobab trees in the savanna. Recurrent fires help create a park effect, with scattered trees set well apart. As it advances, fire only spares mature trees which have thick bark or are fire-resistant.

the sable antelope. They share the area with Lichtenstein hartebeest, or konze, which is very similar to the kongoni.

The carnivores are also well represented, including the hunting dog and the long-eared fox. This park is especially charming. It is partly unexplored and completely wild. It is possible to camp in limited areas.

ZAMBIA

Kafue (18)

Kafue Park covers 8,500 sq. miles (22,000 sq. km) in a vast plateau which is half the size of Switzerland. It is crossed by the tributaries of the Kafue River, running along the eastern boundary. The green and lush vegetation ranges from mixed forest of the "Miombo" type, to bush, to savanna proper. Acacia trees are rare. The landscape is varied because of patches of evergreen forest in the plains. The park hosts sable and roan antelope and eland. It also hosts the rare and unapproachable forest antelope called the bongo, with its striped coat. It also is home for the puku and the bushbuck. Tourist facilities are excellent.

MOZAMBIQUE

Gorongoza (19)

Covering more than 1,400 sq. miles (3,700 sq. km), Gorongoza Park includes various environments. It has subtropical jungle with palm trees and arid bush areas of variably thick and sparse plant life. It has open grasslands and marshes. Attractions of this place include the greater kudu, the sable antelope, and the nyala. Here a pride of lions occupies an abandoned camp. There are leopards and numerous monkeys. There is also a good population of hippopotamuses, crocodile, and waterfowl. Tourist facilities are good, and numerous camps also offer bungalows and rooms.

ZIMBABWE

Wankie (20)

Wankie Park which dates back to 1929, covers more than 5,400 sq. miles (14,000 sq. km). Only some of it can be visited. To the east, the boundaries of the park adjoin the sandy Kalahari Desert. But most of the park has plenty of water and is part of the "Miombo" forest belt. Numerous observation platforms have been built. Some artificial pools have been built so visitors can watch hippopotamuses. The elephant population is about seven thousand head, and the buffalo herds number some four hundred head. In addition, there are plenty of black and white rhinos, which have been reintroduced into the park. Facilities are excellent.

GLOSSARY

adaptation change or adjustment by which a species or individual improves its condition in relationship to its environment.

anthropolgy the study of humans, especially of the physical and cultural characteristics, customs, and social relationships.

arid lacking enough water for things to grow; dry and barren.

atmosphere the gaseous mass surrounding the earth. The atmosphere consists of oxygen, nitrogen, and other gases, and extends to a height of about 22,000 miles (35,000 km).

australopithecene of or relating to a genus of extinct ape-like people from southern Africa that made tools and walked in an upright position.

basin all the land drained by a river and its branches. Water collects near a basin to form lakes.

biogeography the branch of biology that deals with the geographical distribution of plants and animals.

biology the science that deals with the origin, history, physical characteristics, life processes, etc. of plants and animals.

botanist a plant specialist. Botanists study the science of plants, which deals with the life, structure, growth, classification, etc., of a plant or plant group.

camouflage a disguise or concealment of any kind.

carnivore a meat-eating organism such as a predatory mammal, a bird of prey, or an insectivorous plant.

conservation the controlled use and systematic protection of natural resources, such as forests and waterways.

continent one of the principal land masses of the earth. Africa, Antarctica, Asia, Europe, North America, South America, and Australia are continents.

crater a bowl-shaped hole or cavity, such as the mouth of a volcano or the pit formed by a fallen meteor.

deciduous forests forests having trees that shed their leaves at a specific season or stage of growth.

dominant that species of plant or animal which is most numerous in a community, and which has control over the other organisms in its environment.

ecology the relationship between organisms and their environment.

ecosystem a system made up of a community of animals,

plants, and bacteria and its physical and chemical environment.

environment the circumstances or conditions of a plant or animal's surroundings.

erosion natural processes such as weathering, abrasion, and corrosion, by which material is removed from the earth's surface.

evolution a gradual process in which something changes into a different and usually more complex or better form.

extinction the process of destroying or extinguishing.

famine an acute shortage of food.

fauna the animals of a particular region or period. The fauna of any specific place on earth is determined by the animals' ability to adapt to and thrive in the existing environmental conditions.

flora the plants of a specific region or time.

forage food for domestic animals; fodder.

fossil a remnant or trace of an organism of a past geologic age, such as a skeleton or leaf imprint, embedded in some part of the earth's crust.

habitat the area or type of environment in which a person or other organism normally occurs.

herbivore an animal that eats plants. Elephants and deer are herbivores.

humus a brown or black substance resulting from the partial decay of plant and animal matter.

lair a bed or resting place of a wild animal.

lava melted rock that flows from a volcano.

mammal any of a large class of warm-blooded, usually hairy vertebrates whose offspring are fed with milk secreted from special glands in the female.

marsh an area of low-lying flatland, such as swamp or bog. The marsh is a type of borderland between dry ground and water.

microbe a microscopic organism, especially any of the bacteria that cause disease; germ.

migrate to move from one region to another with the change in seasons.

niche the specific space occupied by an organism within its habitat; a small space or hollow.

organism any individual animal or plant having diverse organs and parts that function as a whole to maintain life and its activities.

photosynthesis the process by which chlorophyll-containing cells in green plants convert sunlight into chemical energy and change inorganic into organic compounds.

plateau an elevated and more or less level expanse of land.

precipitation water droplets or ice particles condensed from water vapor in the atmosphere, producing rain or snow that falls to the earth's surface.

predator an animal that lives by preying on others.

prey an animal hunted or killed for food by another.

primate any of an order of mammals, including man, apes, monkeys, lemurs, etc., characterized by flexible hands and feet, each with five digits.

refuge shelter or protection from danger or difficulty; a place of safety.

reptile a cold-blooded vertebrate having lungs, a bony skeleton, and a body covered with scales or horny plates. Snakes and lizards are reptiles.

ruminants cud-chewing animals; grazing animals having a stomach with four chambers.

safari a journey or hunting expedition, especially in Africa.

sanctuary a place of refuge or protection, a place where animals or birds are sheltered for breeding purposes and may not be hunted or trapped.

savanna a treeless plain or a grassland characterized by scattered trees, especially in tropical or subtropical regions having seasonal rains.

scavenger any animal that eats refuse (garbage) and decaying organic matter.

species a distinct kind, sort, variety, or class.

symbiosis the living together of two kinds of organisms, especially where such an association provides benefits or advantages for both.

ungulate of or belonging to a group of animals which have hoofs.

zoologist a specialist in the study of animals; their life, structure, growth, and classification.

INDEX

Aardvark, 27, 45
Aardwolf, 27
Acacia bush, 33, 43
Acacia trees, 8, 18, 20-21, 31, 33, 34, 38, 43, 98, 102
Adders, 104-105
African black kite, 102
Agama, 105
Altruistic behavior, 49, 78-79
Ambroseli National Park, 20-21, 40-41, 76, 87
Andropogon genus, grasses, 18
Anteater, 27
Ants, 32
Areas of natural interest
    Guinea savanna, 108-114
    Kenya, 115-118
    Mozambique, 122
    Tanzania, 118-119
    Zaire, 114-115
    Zambia, 122
    Zimbabwe, 122
Arid-deciduous forest, 16
Aristida genus, grasses, 16, 18
Australopithicus, 83-84

Baboons, 87-90
Bateleur eagle, 102
Bearded vulture, 101
Bellicositermes goliath order termites, 27
Biomass, 25, 27
Birds
    birds of prey, 90, 91, 92, 98, 102, 103
    cooperation with other animals, 91-92
    flightless, 93-95
    migrations of, 91
    scavengers, 98-102
    social habits, weaverbirds, 95, 97
Black-backed jackal, 80, 102
Black cobra, 105
Black-necked cobra, 105
Black rhinoceros, 34, 54
Black-winged kite, 102
Boabab trees, 17, 119, 120-121
Bovidae family, 32, 46
Buffalo, 11, 16, 22, 32, 33, 35, 37, 46, 47, 51, 52, 66, 67, 116-117
Burchell's zebra, 49, 50-51
Bush land, 16

Camouflage, weaverbird, 98
Carnivores
    hunters and scavengers, 62, 63-65
    hunting techniques of, 65-74
    predator-prey system, 61-63
    social behavior of, 74-81

Cattle egret, 91
Cattle plague, 22, 23
Cheetah, 23, 24, 47, 64, 65, 71, 73-74, 75, 79-80
Cleaning birds, 91
Cobras, 105
Coke's kongoni, 45
Common jackal, 80
Communication, guenon monkeys, 85, 87
Cooperation
    among baboons, 87-89
    birds with other animals, 91-92
Cud-chewing herbivores, 31-32

Defensive tactics, herbivores, 47-51
Digestive processes, herbivores, 30-33
Digitaria macroblephora, grass species, 35
Dik-dik, 34, 35, 38, 43
Diluting effect, herd defense, 47
Drought, 15
Duels, male herbivores, 56-59
Dung beetles, 25

Eagles, 90, 91, 98, 102
Eared vulture, 100-101, 102
Ecological niche, 49
Ecological separation, 35, 37
Ecosystems, 7, 8, 9, 10-11, 22, 24, 39, 51, 62, 91, 100, 102
Egyptian geese, 49, 50
Egyptian vulture, 98
Eland, 17, 46, 56
Elephant grass, 16
Elephants, 6-7, 10, 16, 24, 31, 34, 39-42, 52-54, 56, 67, 112-113
Erosion, 13, 16, 33
Euplectes genus, birds, 95

Fires, 16, 18-19, 22
Flamingos, 12
Flightless birds, 93-95
Food availability and social structure, 43-47

Galls, 31
Gambonese adder, 104
Gazelles, 10, 16, 24, 29, 33, 34, 35, 37-38, 45, 46, 47, 48, 59, 62, 66, 69, 70, 71, 74, 80
Geese, 49, 50
Gerenuk, 30, 34, 44
Giraffes, 23, 24, 31, 34, 47, 56-57, 91
Goodall, Jane, 77
Graminaceae family, grasses, 18-19, 29

Grant's gazelle, 10, 16, 29, 38, 45, 63 93
Grass, food value of, 30-33
Grazing, effects of, 29-30
Great bustard, 94-95
Green baboon, 89
Green guenon monkey, 34, 85
Griffon vulture, 63, 99, 100
Grumeti River, 37
Guenon monkeys, 34, 85, 87
Guinea fowl, 96-97
Guinea savanna, areas of natural interest, 108 -114

Harem, 45, 47, 51-52
Hartebeest, 16
Hemmingway, Ernest, 11
Herbivores, 9, 16, 17, 23, 24
  defensive tactics of, 47-51
  digestive processes of, 30-33
  ecological separation and, 35-36
  effects of grazing by, 29-30
  elephant problem, 39-42
  family and solitary life of, 51-56
  feeding habits of, 33-34
  mating habits of, 46, 56-60
  migrations of, 37-39
  social structure, 43-47
Herd defense, 47-51
Hippotamuses, 49, 67, 69
Homo erectus, 83
Homo habilis, 83, 84
Homo sapiens, 83
Hornbill, 92
Hunting dogs, 23, 24, 61, 62, 64-65, 69-71, 77-79
Hunting techniques, carnivores, 65-74
Hyenas, 23, 24, 51, 61, 62, 63, 64, 65, 69, 70, 77-79
Hyparrhenia genus, grasses, 16-17, 19

Impalas, 25, 37, 39, 45, 46, 47, 49, 58-59
Instigation ceremony, hunting dogs, 70-71

Jackals, 61, 62, 80-81, 94, 102, 104-105
Jackson's kongoni, 45
Jaeger, Fritz, 11
Johanson, Donald, 84

Kenya, areas of natural interest, 115-118
Kites, 102
Kob, 34, 44-45
Kongoni, 16, 24, 42, 44, 45, 49, 59, 70

Kopjes, 13, 43
Kruger National Park, 70

Lake Eyasi, 9
Lake Legaja, 9
Lake Nakuru, 12
Lake Rudolf, 26-27
Lake Victoria, 11
Lammergeier, 101
Leakey, Louis and Mary, 83
Leopard, 23, 71-73, 79
Lilac-breasted roller, 91
Lions, 23, 24, 47, 51, 61, 62, 63-69, 74-77, 80
Loita Plains, 9
Long-eared fox, 104
Long-tailed anteater, 27
Loudetia karagensis, grass species, 18
Lucy, 84

Macro-termites, 27
Malimbus genus, birds, 95
Marabou stork, 91
Mara River, 14-15
Martial eagle, 102
Masai herdsmen, 23
Mating season
  great bustard, 94-95
  herbivores, 46, 56-60
Mbalageti River, 37
Microhabitats, 6
Migrations
  bird, 91
  herbivores, 37-39
Mimetism, 98
Mongooses, 27, 92
Monsoons, 13-14
Mozambique, areas of natural interest, 122
Murrain plague, 22

Naabi Hills, 15
Nairobi Snake Park, 105
Ngorongoro Crater, 12, 42, 54, 69
Ngorongoro Mountains, 9
Ngorongoro Plateau, 12
Night adder, 104

Olduvai Gorge, 54, 83, 84, 85, 86-87
Oribi, 44
Oryx, 16, 46, 93
Ostriches, 16, 93-94

Pangolin, 27
Panicum maximum grass, species, 16, 19
Pennisetum genus, grasses, 29
Pennisetum mazianum, grass species, 35
Pennisetum purpureum, grass species, 16, 18
Photosynthesis, 29
Ploceus, family birds, 95
Precambrian time, 13
Predator-prey system, 61-63
Primates
  baboons, 87-90
  guenon monkeys, 85, 87
  origins of, 83-85
Pseudo-territory, 58
Puff adder, 104
Python, 102

Rain forest, 16
Rainy and dry seasons, 13-16
Reptiles, 102, 104-106
Rhinoceroses, 31, 34, 54-57, 67, 91
Rift Valley, 11, 12
River forest, 33
Rumen, 31
Ruminants, 31-32
Ruppell's vulture, 98-99, 100, 101
Ruwenzori Park, 62, 67

Savanna types, 22
Scavengers
  bird, 92, 98-102
  hyenas and jackals, 62, 63-65
Schaller, George Beals, 13, 91
Scrub, 16
Seasonal grazing, 33
Secretary bird, 102, 103
Semideciduous forest, 16
Serval, 104
Side-striped jackal, 80
Social behavior
  carnivores, 74-81
  herbivores, 43-47, 51-56
  weaverbirds, 95, 97
Spotted hyena, 63
Steppe, 33
Storks, 94
Stotting, 48
Sukuma tribe, 9
Symbiosis, 30-31, 91-92

Tana River, 45
Tanzania, areas of natural interest, 118-119, 122
Tawny eagle, 90, 91, 98, 102
Termites, 25-27, 92
Themeda genus, grasses, 19, 29
Themeda triandra, grass species, 16, 37
Thomson's gazelle, 10, 24, 29, 33, 34, 35, 37-38, 45, 48, 49, 59, 66, 69, 70, 71

Three-bearded grass, 16
Topi, 16, 24, 34, 35, 37, 44, 45, 49, 70
Trees, 16-18
Tsavo Park, 30, 34, 116-117
Tubulidentates order anteaters, 27
Tumuli, 27

Vegetation sequence, 16
Volcanoes, 12-13
Vultures, 62, 63, 80, 91, 94, 96-102

Warrior termite, 27
Warthogs, 24, 48-49, 75
Waterbuck, 31-32, 34
Weaverbirds, 95, 97
White-backed vulture, 98
White-headed vulture, 98
White rhinoceros, 54
Wildebeests, 10, 11, 14-16, 22-23,
    24, 29, 31, 35, 37, 39, 44, 45-46,
    47, 51, 55-57, 62, 65-66, 69, 70

Zaire, areas of natural interest,
    114-115
Zambia, areas of natural interest,
    122
Zebra, 10, 22, 31, 33, 35, 37-38, 49,
    50-52, 62, 66, 69, 70
Zimbabwe, areas of natural
    interest, 122

CREDITS

MAPS AND DRAWINGS. G.Vaccaro, Cologna Veneta (VR). **PHOTOGRAPHS. M.L.Banfi,** Vimercate (MI): 70, 80-81, 85, 96-97. **G.Bardelli,** Udine. 6-7, 23, 67, 74, 75, 92. **L.Beani,** Florence: 48-49, 50-51, 105, 106. **A.Borroni,** Milan; M.Mairani, 63, 76-77. **Diamonde,** Turin: R. Sacco, 111. **C. Giacoma,** Turin: 14-15. **Marka Graphic,** Milan: 36-37. **Panda Photo,** Rome: P.Harris, 72-73, 116-117; M. Paluan, 82; G.Petretti, 71; W.Rossi, 30, 31, 38, 95, 99, 104; F.Vollmar, 101; A.Zocchi, 17m 32. **L.Ricciarini,** Milan: D.Biglino, 93; N.Cirani, 8, 20-21, 26-27, 54-55, 57, 60, 86-87, 89; Archivo, 46-47, 64-65; A.P.Rossi, 90, 103; **M.P. Stradella,** Milan: 11. **Foto Studio Wemme,** Brescia: W. Pescara, 40-41, 112-113; M.Pezzotta, 12-13, 28, 42, 120-121.

REFERENCE--NOT TO BE
TAKEN FROM THIS ROOM

DATE DUE

DATE DUE

```
574.5      Beani, Laura
Bea
           The African
           Savannah
```